BOMBARDIER BEETLES
AND FEVER TREES

by the same author

Chemical Communication:
The Language of Pheromones

BOMBARDIER BEETLES AND FEVER TREES

*A Close-up Look at Chemical Warfare
and Signals in Animals and Plants*

William Agosta

HELIX BOOKS

ADDISON-WESLEY PUBLISHING COMPANY, INC.

Reading, Massachusetts   Menlo Park, California   New York
Don Mills, Ontario   Harlow, England   Amsterdam
Bonn   Sydney   Singapore   Tokyo   Madrid   San Juan
Paris   Seoul   Milan   Mexico City   Taipei

Library of Congress Cataloging-in-Publication Data

Agosta, William C.
   Bombardier beetles and fever trees : a close-up look at chemical warfare and signals in animals and plants / William Agosta.
       p. cm.
   Includes bibliographical references and index.
   ISBN 0-201-62658-6
   ISBN 0-201-15497-8 Pbk.
   1. Chemical ecology. I. Title.
   QH541.15.C44A38 1995
   574.5—dc20   95–9533
                                                                      CIP

Cover design by Lynne Reed
Text design by Joyce Weston
Set in 10.5-point Janson by Carol Woolverton Studio

1 2 3 4 5 6 7 8 9-MA-0100999897
First paperback printing, April 1997

CONTENTS

LIST OF ILLUSTRATIONS

1

NATURE'S STOREHOUSE
OF CHEMICALS

*Y*EARS AGO when I was a boy in Texas, I played outdoors much of the year and spent summers on a farm surrounded by fields of cotton and corn. The ordinary plants and animals of that world were probably not much different from many others, but they were certainly more varied than those I meet nowadays, living and working in the middle of a metropolis. I am reminded of some of those plants and animals from earlier years in connection with the exploration we are about to undertake. How these creatures are involved in our story will be clear shortly, but let me first recall a few of them.

There were bees and yellow jackets that stung when they were annoyed, and also stinging nettles and poison ivy that provoked welts and itchy rashes. There were fireflies that flashed their strange greenish lights throughout warm summer evenings. Colorful butterflies darted among bright, fragrant flowers, and green stinkbugs pursued their own affairs here and there. Sharp-tasting mint grew in the cool shade behind the house, and milkweed, filled with sticky sap, thrived in weedy places. Spiders hung their lovely webs on bushes or spread them in the grass. Skunks were rare, but when one was hit on a country road, its lingering scent spread far and wide.

These were everyday creatures in that natural world, and I thought little about them. Even if I had thought about them, I could not have appreciated the debt they all owe to their own special chemicals. I did not realize that the bee's sting, the flower's fragrance, and the spider's web all derive from specific chemicals that are part of those creatures' daily lives. Chemicals provide the mint's zesty flavor and the milkweed's sticky milk, and in doing so, they serve these plants in important ways. I did know that bees smell unlike skunks, and that skunks do not sting, so I could have understood that creatures use their special chemicals in many different ways.

I would have been surprised to learn what varied things living organisms do with their chemicals. Countless creatures in all corners of the earth employ chemicals to deter predators, attract mates, and seek out prey. They send one another warning signals and broadcast calls for help. They create protective camouflage, make glue, lay trails, and poison their enemies.

To carry out these many tasks, creatures have developed chemicals that have distinctive, sometimes unique, characteristics. Only particular chemical compounds can endow the skunk with its foul spray or lend flowers their attractive colors. Spiders construct their webs using chemicals that have truly singular properties. Thousands of chemicals with specific characteristics are critically important to the creatures that make and use them. This vast collection of chemical compounds produced by particular organisms is what I call "nature's storehouse of chemicals," or simply, "natural chemicals." The ordinary plants and animals of my childhood have now become the foundation of this story of living creatures and how they use their chemicals.

Another part of this story is about how we humans make use of these natural chemicals for our own purposes. Chemicals from nature have been a part of human civilizations ever since our early ancestors began exploiting natural compounds to improve and enrich their own lives. This usage began long ago, long before there was the slightest concept of chemistry and biology as sciences, and these people had no notion that they were making use of chemicals. They had simply learned that plants and animals could supply them with beneficial commodities, and they were taking advantage of their discoveries.

The variety of these discoveries reflects the familiarity many earlier civilizations had with the world around them. South American Indians tipped their spears and arrows with poisons they took from plants and frog skins. Roman soldiers stuffed spider webs into sword wounds to stanch bleeding and prevent infection. Aztec priests consumed holy mushrooms containing mind-altering chemicals in order to enter into the world of the gods. The opium poppy offered medicine and magic to peoples around the Mediterranean, centuries before the rise of Greece or Rome. Half a world away, the rulers of ancient China guarded the secret of the silkworm and cherished silk as a priceless gift from heaven. These age-old contributions of nature's chemicals to human welfare have modern counterparts in many of our specialized drugs, antibiotics, and insecticides.

We continue today to extract an array of invaluable products from nature. As the array of products has expanded, the natural storehouse of chemicals has had an increasing impact on our daily lives, and it comes more and more to public attention. Topics such as pesticides and pollution, cancer and diet, new drugs from nature, and human sex attractants, all are familiar to millions. They make frequent headlines and are featured on the nightly television news. Major newspapers publish science sections that regularly consider such topics seriously and critically. These subjects all have important associations with natural chemicals. We shall pursue these connections in detail as part of our story: here at the outset, we can summarize some of them briefly.

The complex problem raised by pesticides and environmental pollution is now well known to millions of people. Pesticides and their polluting effects first entered American public consciousness with the appearance of Rachel Carson's *Silent Spring* in 1962. Even at that time, the pesticide problem was already a virtually intractable mix of science, politics, and economics, so it is not surprising that, after more than thirty years of research and discussion, we have still failed to fashion an acceptable solution. While the thirty-year effort has certainly not solved the problem, it has made us much more aware of the environmental price we pay for agricultural productivity that was unimaginable in an earlier era.

Despite much concern and anguish, chemical compounds that contribute to environmental pollution still completely dominate the pesticide market. We dispense half a million tons of these pesticides each year to control crop pests in the United States and to make our farmland phenomenally productive. The chemical runoff from these pesticides carries pollutants from farmers' fields into rivers and lakes, and eventually into coastal waters. All along the way, pollutants disrupt the balance of living systems and

3

degrade the environment. In the mid-1980s, the destructive side effects of pesticides were costing the United States about $800 million annually. Meanwhile, significant crop pests become resistant to one pesticide after another. Increased public awareness of environmental issues has led to many encouraging improvements, but substantial problems remain. In September 1994, the federal government moved to reduce the use of chemical pesticides in the United States, noting their harmful consequences to both the environment and the health of farm workers.

The connection of the pollution problem to our story is a direct one. Natural chemicals offer environmentally safer approaches to controlling pests than those now in use. Agricultural scientists are actively studying several kinds of chemicals from nature's storehouse, including chemicals that natural organisms use to carry messages, chemicals that plants themselves use to deter predators, and insecticides that are made by bacteria. Each of these substances offers possible alternatives to chemical pesticides, and the general goal is to make these natural approaches truly practical and competitive. Some studies have already led to commercially available products for pest control.

In quite a different direction, nutritionists have emphasized the substantial role of natural chemicals in human health. As evidence has mounted linking diet with about 30 percent of all cancers, dietary specialists have pressed us with growing urgency to eat fruits and vegetables. Nutritional and epidemiological studies point to certain plant chemicals that have specific anticancer roles. Chemicals that protect broccoli from insect predators also could help prevent cancer in humans. A compound in soybeans appears to combat human cancer by slowing the formation of the new blood vessels necessary for tumor growth. As nutritionists point out, preventing cancer is preferable to treating it. Eating fruits and vegetables has few disagreeable consequences, but treating cancer with radiation, surgery, and chemotherapy often has extremely unpleasant side effects.

Nor is it only against cancer that fruits and vegetables seem to be effective. In November 1994, *The New York Times* commented editorially on the advantages of our eating yellow and green vegetables to protect against a common form of blindness and against heart disease as well. To get the most from nature's chemicals, the experts say we should eat five servings of fruits and vegetables every day. It seems that Mom was right all along!

Topics closely related to dietary chemicals include some forms of alternative medicine and the search for new drugs. Instead of eating plants to profit from their chemicals, we can take compounds out of plants and turn them into drugs. Plant-derived drugs have been the mainstay of traditional medicine in many cultures for centuries, and they occupy an important place in modern Western medical practice as well. In the early 1990s, six of the twenty most widely prescribed drugs in the United States came from plants, and serious efforts are currently underway to identify more new drugs from natural sources. Under programs sponsored by such institutions as the National Institutes of Health, the New York Botanical Garden, and several large American pharmaceutical companies, scientists are scouring the globe for possibilities. One group is examining traditional Asian drugs, while another is interviewing local healers on small Pacific islands and testing samples of their medicines. Other teams are screening North and Central American plant species for novel chemicals with biological activity.

The goal of these programs is to identify leads for new drugs that can be incorporated into tomorrow's clinical medicine. The search concentrates primarily on plants, because plants furnish most of the traditional medications employed around the world. However, one of the most promising discoveries of the 1990s is a new chemical compound derived from dogfish sharks (*Squalus acanthias*) that kills parasites, fungi, and bacteria. This chemical helps prevent infections in wounded sharks, and someday it may do the same for us.

1.1 Sharks effectively resist infection, although many biologists believe that their immune system is poorly developed. Their resistance may depend on an antimicrobial compound recently discovered in dogfish sharks.

Another popular biochemical topic of the 1990s is human pheromones. Many people are now aware that creatures everywhere exchange chemical messages with other members of their species. The chemicals that carry these messages are called pheromones, and the pheromones most widely discussed in the popular press are sex attractants. Numerous insects, mammals, and other creatures broadcast attractants that bring the sexes together and so facilitate mating.

Popular discussions of pheromones inevitably lead to exciting and titillating speculation about sex attractants in humans. Can our sense of smell guide us to fulfillment of our dreams? Could love and happiness be just a sniff away? Recent years have brought tantalizing developments regarding human pheromones, both in the laboratory and in the marketplace. Pheromones and other message-bearing compounds occupy a prominent place in our storehouse of chemicals, and we shall have much more to say about them.

The chemicals connected with these familiar topics, along with many other natural compounds, are a rich source of diversity and complexity in the biological world, and they generate a host of research problems that many scientists are eager to probe. These problems cover a wide range of special interests. What compounds does an organism use? What functions do the compounds serve and how do they work? How are the compounds made by the organism? What effect do the compounds have on other organisms, and how do these other organisms respond?

Much research of this sort falls within an interdisciplinary area generally known as chemical ecology. The name reflects the fact that ecology is the science devoted to relationships between organisms and their environments. Organisms use their special chemicals in the course of interacting both with other organisms and with their physical surroundings, that is, they use these chemicals in interacting with their environments. Thus the designation, "chemical ecology."

One of the most appealing aspects of chemical ecology is its broad accessibility. Much of its subject matter illustrates directly and plainly the interconnectedness and marvelous complexity of nature. Its raw material is meaningful to anyone fascinated by the natural world. Unlike some other sciences, many of its research problems are easy for everyone to comprehend on a general level, even if the solutions require the combined efforts of experts in several fields. For example, how and why is it that, after one bee has stung a particular spot, other bees excitedly sting the same place? Why is it that certain plants attract many fewer insect predators to their leaves and flowers than do their neighbors, and those few predators that do come hide themselves from view while they feed? How do Antarctic fishes avoid freezing as they swim in their ice-filled ocean, although a mid-Atlantic fish freezes immediately in Antarctic waters?

No formal scientific training or elaborate equipment is necessary to raise these questions, and there are endless questions of this sort. The question about bees was first raised early in the seventeenth century, yet it was satisfactorily answered only in recent decades. Such questions initiate cascades of riddles for curious investgators. As scientists solve the riddles, the extraordinarily complex chemical interactions among living creatures are slowly becoming apparent. The story we are about to explore comes largely from the accumulating answers to such riddles about our natural world.

*B*EFORE delving into this world, perhaps we should briefly examine what we mean by "chemicals," "chemical compounds," or simply "compounds." We have been talking very freely about them, and we will use these terms interchangeably throughout our explorations. The chemical compounds that figure in our story are substances made up of what chemists call organic molecules. A molecule is a group of atoms joined together in a specific three-dimensional arrangement or structure. "Organic" means only that these molecules contain carbon atoms, a normal property of molecules found in living systems. (Hundreds of thousands of other organic molecules never found in nature have been prepared in the chemist's laboratory.) Different molecules may consist of different numbers of atoms, from two to thousands, so molecules may be of various sizes and degrees of complexity. Two molecules having exactly the same number and the same kinds of atoms may have quite different structures, owing to the diverse ways in which atoms can join together. The structure makes the compound what it is; that is, the structure is responsible for the compound's properties and chemical behavior. Like everything else made of atoms, all molecules are real physical objects, though even large molecules are still very small objects.

This basic understanding about chemical structure will be helpful for our story. Occasionally, for example, we shall mention transforming one structure into another. Another significant idea will be that two different molecules may have complementary shapes so that they fit snugly together. We shall not need, however, to go into further detail about chemical structure.

For convenience, we can think of the chemicals in nature's storehouse as belonging to three groups. There are agents of chemical warfare, such as the skunk's spray or poison ivy's irritant. There are compounds that play a central role in lifestyles, such as those responsible for the firefly's light or those from which a spider constructs its web. Finally, there are chemicals that convey messages, such as sex-attractant pheromones and the scents of flowers. These three groups furnish a convenient framework, but they are only loose classifications. Their limitations will soon be apparent, and we shall find that many compounds and their biological roles can fit quite comfortably into more than one of the three categories. The skunk's spray is a fine chemical warfare agent, but at the same time it is also a defining element of the skunk's lifestyle that delivers a clear message to other creatures.

We cannot include in this book all that is known about these natural chemicals. There is simply too much information. However, we can sample organisms and the natural chemicals associated with them at many different points throughout nature, looking at widely diverse creatures and chemicals with various properties and uses. In this way we can illustrate the enormous variety of interactions among organisms that chemicals provide. As we progress, it should become clear that these natural chemicals literally unite the biological world. At the same time, we shall also see how these chemicals affect our own lives and what they can do for us.

One final point before we begin: each organism's scientific name, that is, the Latin name of its genus and species, appears when the organism is first mentioned, as for the dogfish shark referred to earlier in this chapter. The "Note to the Reader" that follows chapter 9 contains a brief discussion of biological classification that may be helpful. It also explains the scientific units of measurement used throughout this book.

2

CHEMICAL DEFENSES
IN THE PLANT WORLD

*P*LANTS MAY seem to be passive creatures, but they defend themselves vigorously against their enemies, frequently by means of chemical warfare. Many plants synthesize or deploy chemicals specifically in response to an attack by a predator, often driving the predator away or even destroying it. Other plants, such as stinging nettles and poison ivy, synthesize defensive chemicals in the course of their normal metabolism and store them in various tissues, creating a chemical stockpile against future assault. Interactions between plants and herbivores (organisms that feed on plants) are complex, and biologists often find it difficult to elucidate the exact function and the effectiveness of these defensive compounds. The presence of a defensive chemical may be only one of several factors protecting a plant from predation. Plants contain chemicals with unusual and even remarkable properties, but in many cases we do not yet wholly understand the defense systems within which these chemicals operate.

Yet, we do know that defensive chemicals act in several different ways. Many are toxic to a wide variety of creatures. Others are toxic only to certain organisms or only under certain circumstances. There are compounds employed to enlist the help of other species in repulsing enemies, and ones that weaken predators and render them more susceptible to disease. Some chemicals turn into a gooey mass and entrap a predator physically, and others mimic or block the action of a predator's hormones, interfering with its growth and development.

The techniques that plants use to defend themselves have attracted intense scientific attention for years, and scientists now investigate plant defenses with all the methods of modern molecular biology and chemistry. There is enormous practical interest here, largely because agriculture holds such a critical position in human welfare. A better understanding of how plants defend themselves can lead to more effective methods of producing and protecting our crops, a critical goal throughout the world. As such, this research offers hope in Asia and Africa for alleviating starvation and malnutrition. In Europe and North America, it can reduce the enormous environmental price paid for the heavy use of agricultural pesticides.

There is a huge mass of information on defensive chemicals in plants, and we can only sample it here. The following examples illustrate the variety of chemical defenses and how they fit into the world of plant and predator.

BEETLES AND BITTER MELONS

One of the most striking characteristics of plant chemical defenses is their dynamic nature: they change and develop over time. The predators that attack plants are mainly insects, and these two groups have a long history of life together. Insects have been eating plants for perhaps 300 million years, and yet plants still flourish, so they have defended themselves successfully against this assault. At the same time, insects have continued to eat plants, so they have successfully overcome or circumvented these plant defenses. Neither group wipes out the other; each continues to thrive. Both plant defense and predator response are always changing, and plant and predator evolve together. Sometimes the changes are slow and become apparent only after careful study, and sometimes they take place quickly. Rapid adaptation by insects is perhaps most familiar as a response not to plant defenses but to attack by human beings. Agricultural pests can develop resistance to a pesticide so rapidly that the amount of pesticide required to control the same number of pests increases from year to year.

Changing plant defenses and herbivore responses often give rise to complications and consequences that we would never foresee. Consider, for example, the very effective defense chemicals of the gourd family and how a

10

group of beetles has turned these chemicals to its own advantage. Roughly nine hundred species of gourds, squashes, and melons make up a single family of plants, found primarily in tropical or subtropical climates. At least one hundred of these species contain bitter substances sometime during their development. Chemists have isolated and identified about ten of these closely related bitter compounds, which are called cucurbiticins (derived from the plant family's botanical name, Cucurbitaceae), and to human taste they are the bitterest substances known. Tests demonstrate that humans can reliably detect as little as 1 part per billion (thousand million) of one of these cucurbiticins dissolved in water. This is equivalent to 1 microgram per liter of water. To appreciate how small that amount is, consider that typical grains of table salt weigh roughly 100 to 500 micrograms each. The amount of water ingested in this test is probably only 1 or 2 milliliters, so human beings can detect roughly one thousandth of the amount of material dissolved in a liter, or roughly 1 nanogram of the cucurbiticin.

Not only are these compounds extremely bitter, they are also poisonous, and domestic livestock occasionally die from overfeeding on bitter gourds and squashes. Because synthesizing and storing these bitter toxins is expensive for the gourds, these activities must justify themselves. The obvious justification is that the unpleasant taste and toxicity of the cucurbiticins deter predation. Most herbivores, both vertebrate and invertebrate, stay away from the gourds and melons that contain these bitter principles.

However, there is a tribe of over fifteen hundred species of beetles that specializes in feeding on these plants. This group includes three cucumber pests known as the spotted, banded, and striped cucumber beetles (*Diabrotica undecimpunctata howardi*, *Diabrotica balteata*, and *Acalymma vittata*, respectively). Entomologists (scientists who study insects) have recorded the feeding habits of only a small portion of the species in this tribe, but 80 percent of the ones examined feed on plants in the gourd family. The beetles are not merely undeterred by the cucurbiticins: they are positively attracted to these compounds. They have specialized receptors that detect and respond to nanogram amounts of the cucurbiticins, displaying about the same sensitivity as humans. The beetles can detect a cucurbiticin-containing gourd plant at a distance of several meters, particularly if it is bruised or wounded. When they detect such a plant, they speed toward it. When they reach the plant, they begin feeding. The insects become literally compulsive

11

about eating the bitter compounds. When an investigator offered them small grains of sand coated with a cucurbiticin in a laboratory experiment, the beetles ate the sand as though they were starving. For these specialized insects, the gourds' defense compounds have become an irresistible attractant and feeding signal.

WHY INSECTS AVOID ST. JOHN'S-WORT

There are many other plant toxins with surprising properties. One of the oddest is that of St. John's-wort, or Klamath weed (*Hypericum perforatum*), a common weed that grows along roads or at the edge of meadows throughout much of the United States. This European plant was transported to the New World in colonial times and has spread across the continent with great success, in part because animals generally leave it alone. If you examine the leaves of St. John's-wort and look into its bright yellow flowers, you will discover very few insects feeding on them. Most insects reject this weed, as do livestock.

A few North American insects have adopted St. John's-wort as a food source. Most of these are butterfly or moth larvae that hide themselves while they feed. Some of these larvae roll or fold a leaf and bind it with silk to cover themselves. Others, called leaf tiers ("tie-ers"), sew leaves together to form a shelter. Still other species bore into a stem and devour the plant from within. One group, the leaf miners, has small flat larvae that actually live inside leaves. They eat out tiny tunnels, ingesting the middle layers of leaf tissue without disrupting the top and bottom of the leaf. All of these hidden feeders forage on St. John's-wort with little competition, because most animals that feed in the open shun the plant.

Hidden feeders of St. John's-wort must have some advantage over other predators. Obviously, by concealing themselves, they are shielded from their own enemies. However, there is even more going on here: hidden feeders are also shielded from direct sunlight. Avoiding enemies is always desirable, but for larvae foraging on St. John's-wort, avoiding sunlight turns out to be absolutely vital.

This phenomenon may be surprising, but consider a feeding experiment with one of the hidden feeders, the larvae of the moth *Platynota flavedana*. These larvae are leaf tiers that ordinarily thrive on St. John's-wort. Entomologists at the University of Illinois brought the larvae into the laboratory, kept them in full light, and fed them a standard diet supplemented with St. John's-wort leaves. Under laboratory conditions, the larvae could not sew their leaf shelters, but were forced to feed in the open. When the larvae could not avoid direct light as they ate, St. John's-wort killed them. Control experiments showed that if the larvae were shielded from full light in the laboratory, or if the leaves were omitted from their diet, they survived nicely.

The experiment demonstrates that the toxicity of St. John's-wort is dependent on the presence of light. The plant is poisonous when eaten in the light but safe to eat in the dark. St. John's-wort is one of many plants that contain a toxic defensive chemical whose effects are triggered by light. Such light-triggered defense chemicals are called phototoxins. In St. John's-wort, the phototoxic compound is a chemical called hypericin. Hypericin is made and stored in glands in the leaves, flowers, and stems of St. John's-wort, and is ingested by any feeding predator. Along with ingested nutrients, hypericin is absorbed from the predator's gut and distributed throughout its body. In this way, molecules of hypericin reach the outer surface of the organism, where they absorb some of the light that falls there.

Before we can discuss what happens to the predator, we should look at what happens to hypericin when it or any other molecule absorbs light. Absorbing light means absorbing energy. Like a person who has just caught a very hot potato, the molecule wants to get rid of this excess energy quickly. There are several ways to dispose of excess energy, but only two of them are important to us here. The first is that a molecule may simply pass the energy on to another molecule. One molecule thus solves its problem but creates the same problem for another. The second way of getting rid of energy is for a molecule to form new chemical bonds with another molecule. This uses up the excess energy, but both initial molecules are destroyed in the process, and one or more new molecules are created.

Now we can consider what happens when an ingested hypericin molecule absorbs light. Investigators have not yet established the details, but it is very likely that the hypericin molecule quickly disposes of its excess energy by transferring it to an oxygen molecule. Oxygen is plentiful throughout the organism, including the surface areas where a hypericin molecule can absorb sunlight. A nearby oxygen molecule readily accepts energy from hypericin, and the hypericin molecule returns to its normal state, capable of

13

absorbing light again and repeating the cycle. The oxygen molecule now has excess energy, which it gets rid of by forming new chemical bonds. The oxygen will react with almost any other molecule that is close at hand, effectively destroying it. Often a reactive oxygen molecule reacts with and destroys molecules of the proteins or nucleic acids that play vital roles in the organism. If oxygen reacts with enough of these molecules, the organism dies. The chain of events that starts with hypericin absorbing light can kill a predator from within, one molecule at a time.

We can see that hypericin differs from ordinary poisons in an important way. Ordinary poisons interact directly with some vital mechanism in an organism, disrupt it, and cause harm. Hypericin, on the other hand, is dangerous only after it absorbs light. Without excess energy, hypericin is innocuous. By keeping out of the light, hidden feeders inhibit the photoactivation of hypericin and escape its deadly consequences. In time, they excrete hypericin without adverse effects.

The unpleasant consequences of foraging on St. John's-wort and its close relatives are not limited to insects. Sheep, cattle, and horses will accept *Hypericum* species as food if other forage is scarce. When these animals ingest hypericin, unpigmented skin without hair becomes particularly sensitive to light, so that in livestock the effects of hypericin poisoning are greatest around the mouth, nose, and ears. Dark animals with white spots show the greatest effects on the spots, because dark areas contain skin pigments that absorb much of the light and prevent it from reaching hypericin. In livestock, mild cases of poisoning cause noticeable skin irritation. Severe poisoning leads to blistering, destruction of red blood cells, peculiar body movements, and occasionally death. Serious damage around the mouth sometimes inhibits drinking and feeding, so that poisoned animals may also die of thirst or starvation.

Because of these dangers, hypericin poisoning has long attracted the attention of farmers and ranchers. An eighteenth-century Italian writer reported that after *Hypericum crispum*, a Mediterranean relative of St. John's-wort, had poisoned white sheep around the southern town of Taranto, shepherds in that region kept only black sheep, which resisted the poison. Their dark pigmentation allows black sheep to resist hypericin, but the actual reason why black sheep survived was unknown in the eighteenth century. Their resistance was explained by the then-prevalent belief that black animals were generally hardier than white ones.

In the western United States, St. John's-wort, known locally as Klamath weed, has caused problems for ranchers, particularly because it can displace useful forage plants. In the mid-1940s, when over 1 million acres of western grazing land was covered with the weed, government entomologists mounted a campaign to eradicate it in California. They introduced a little European beetle, *Chrysolina quadrigemina*, whose larvae feed on *Hypericum* species and indeed proved to have a voracious appetite for St. John's-wort. The beetle larvae eat only *Hypericum* species and apparently suffer no adverse side effects. Within four years, the beetle had largely obliterated the weed, and grass began to return to the range. Agricultural scientists have introduced related beetles into Australia and Canada to control St. John's-wort.

Both larvae and adults of several *Chrysolina* species feed only on St. John's-wort and its relatives. The adults of at least one species are actually attracted to hypericin. A British neurophysiologist, C. J. C. Rees, has shown that the adult forms of *Chrysolina brunsvicensis* detect hypericin by means of chemical receptors on their lower legs. Guided by signals from these receptors, the beetles seek out and feed on the upper part of the plant where the toxin is concentrated. Insects with such feeding habits obviously require some means of guarding against hypericin poisoning. Larval and adult beetles accomplish this in different ways. The larvae protect themselves by avoiding light. Young larvae seclude themselves in leaf buds while they feed, and older larvae forage only at dawn and then bury themselves in the soil for the rest of the day. The adult beetles, however, enjoy the sun. They forage in full sunlight and then bask on the leaves, evidently unscathed by the hypericin they have consumed. In fact, *Chrysolina* species have most effectively controlled St. John's-wort in sunny fields rather than shady locations.

The adult beetles' resistance to the phototoxin has a simple explanation. Unlike the soft-bodied larvae, the adults are covered with a dense outer layer, or cuticle. Only about 0.2 percent of the sunlight that could activate hypericin passes through their opaque cuticle. The phototoxin remains inert in their bodies, and these insects ingest it with impunity.

In addition to many species of *Hypericum* that produce hypericin, other plants from at least fifteen different families contain phototoxins. Many of these compounds activate oxygen in much the same way that hypericin does. Successful insect predators of these phototoxic plants have developed two chemical defenses beyond simple light avoidance and opaque coverings.

Some species have specific enzymes that intercept activated oxygen. These enzymes are large protein molecules that promptly deactivate reactive oxygen molecules before they can attack and destroy other molecules. There are several enzymes of this sort, each working chemically in a slightly different way. All of them reduce the ravages of activated oxygen with minimal damage to the enzymes themselves. In some insects, the ingestion of a phototoxin elicits a rapid synthesis of these enzymes, making them available in large amounts just when needed.

The second type of chemical defense also uses molecules with special properties, but in this case they are small molecules that are already present in the organism for other purposes. These molecules, called antioxidants, act as traps for active oxygen. The antioxidant molecules are both effective acceptors of excess energy and also attractive targets for chemical attack by oxygen. They themselves may be removed without ill effect to the organism.

Two widespread compounds that work in this way in insects are vitamin E and beta-carotene, a common red-orange pigment found in many organisms, including carrots. Biologists at the University of Ottawa recently demonstrated that the greater the concentration of antioxidant in the tobacco hornworm (*Manduca sexta*), the less susceptible the insect was to phototoxin poisoning. These investigators also found that larval *Chrysolina hyperici*, which feed only on St. John's-wort, normally contain an unusually large amount of beta-carotene. In addition to light avoidance, *Chrysolina* larvae may safeguard themselves by maintaining a high concentration of beta-carotene in their bodies.

Finally, in view of its photoactivity, it is significant that St. John's-wort has a long history as a medicinal herb. It has found use for complaints as varied as varicose veins, gastritis, and menstrual disorders. In medieval Europe, plants gathered on June 24, St. John's Day, were thought to possess magical properties, and this belief led to the plant's peculiar name. Phototoxicity certainly gives St. John's-wort exceptional chemical and biological properties, and an extract of the plant has the striking, deep red color of hypericin. It may be that these characteristics led to St. John's-wort's reputation as a medicinal, and in fact, hypericin has shown promise in recent studies as a new therapeutic and diagnostic agent.

IMITATING A PREDATOR'S OWN SIGNALS

Karel Sláma had raised tens of thousands of European linden bugs (*Pyrrhocoris apterus*) in his laboratory in Prague. Each bug had grown through five larval stages, or instars, molting after each stage and emerging each time slightly larger than before. The fifth instar then molted to produce the winged, sexually mature, adult form of the bug. The growing larvae had no wings but otherwise looked much like the adults. (This form of insect development is called gradual metamorphosis. Some insects undergo what is called complete metamorphosis, in which an active wormlike larva passes into an immobile pupa and then emerges as a winged adult. Both kinds of development are common in insects.)

However, when Sláma tried to raise linden bugs in Cambridge, Massachusetts, they did not reach maturity. Sláma had come to the Harvard Biological Laboratories in the mid-1960s to work with Professor Carroll M. Williams. As soon as Sláma had set up his laboratory, he began cultivating a batch of linden bugs for new studies. He grew the bugs in small laboratory containers called petri dishes and fed them on linden seeds, their natural food, just as he had in Prague. Their development began normally, but to Sláma's surprise, the fifth-instar larvae failed to turn into adults. Instead, they molted into an exceptional sixth larval stage, and five days later, about half of them molted again into a seventh larval instar. Since insects grow with each molting, these abnormal larvae were giants. Sláma's seventh-instar larvae were 14 millimeters long, although the length of a normal adult linden bug is about 10 millimeters. All the abnormal larvae died without reaching maturity.

Sláma and Williams naturally assumed that some unsuspected difference in growing conditions between Prague and Cambridge must be responsible for the bugs' arrested development and death. Sláma made a detailed comparison of his two laboratories and discovered fifteen differences to consider. He focussed attention first on the fact that silkworms were also growing in the Harvard laboratories. In a laboratory, airborne signals from one species of insect can disrupt the development of another species housed nearby, so the silkworms were likely culprits. Sláma removed the silkworms, but this had no effect on the linden bugs. He turned to other possibilities. He switched the bugs' water, the water vials, the petri dishes, and the linden seeds used for food. None of these changes made any difference.

Finally, Sláma turned his attention to the paper liner in each petri dish. In Prague he had put a piece of laboratory filter paper in the bottom of each dish, but in Cambridge he was using a piece of ordinary paper towel. Sláma replaced the towelling with filter paper, and suddenly the bugs resumed normal development. Fifth-instar larvae molted into proper winged adults. Sláma and Williams were forced to conclude that something in an everyday paper towel had interrupted the metamorphosis of their linden bugs. They called this something the "paper factor," and Sláma began to investigate whether it was present in other paper goods. He examined all sorts of common paper towels, table napkins, facial tissues, wax paper, and laboratory tissues, from a variety of manufacturers. To his surprise, all of these products

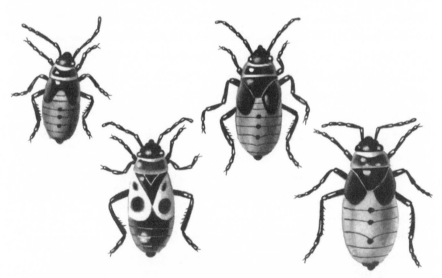

2.1 *The fifth-stage linden bug larva (left) normally turns into a winged adult (second from left). Under the influence of the paper factor, the fifth-stage larva turns instead into a larger sixth-stage larva (second from right), which sometimes goes on to a monstrous seventh stage (right).*

contained the paper factor. He turned to newspapers, and some geographic distinctions appeared. *The New York Times, The Boston Globe*, and *The Wall Street Journal* were even more active than paper towels, but the *Times* of London had no effect on the linden bugs. *Science*, published in the United States, was active, but *Nature*, a British publication, was inert. These peculiarities led Sláma to compare European, North American, and Japanese paper goods. He discovered that only North American paper products contained the paper factor. Japanese newspapers, which were highly active, appeared briefly to be an exception, but then Sláma learned that Japan imported its newsprint from Canada.

With the paper factor so widespread, Sláma and Williams presumed that something in North American paper pulp was responsible. They turned to the wood of American pulp trees and found high activity in extracts of several of them, including the balsam fir (*Abies balsamea*), which is a principal North American pulp tree. Further investigation finally established that this was indeed the source of the paper factor. Soon other investigators were able to isolate the paper factor and show that it is a natural constituent of the balsam fir, a chemical compound that became known as juvabione. It was the tiny amount of juvabione in small pieces of an American paper towel that had arrested the maturation of Sláma's linden bugs.

To understand how paper towelling could interfere with a bug's development, we must look a bit more closely at metamorphosis. A linden bug larva has a hard cuticle. As the bug grows, it must continually molt, or shed its old cuticle, and form a new, larger one to have room for further growth. Molting, like other aspects of growth and development in all creatures, is controlled by internal chemical messengers, or hormones.

In general, a hormone is synthesized in one part of the body and is then carried to distant sites by the circulatory system. At the distant site, a protein molecule called a receptor recognizes the hormone molecule by its shape. The two molecules fit together like lock and key, and the hormone binds to the protein. Recognition of the hormone by its receptor triggers some change in the organism, and this change is the "message" that the hormone delivers. In insects, a gland in the head produces a hormone called juvenile hormone (JH) that helps control development. In linden bugs, the gland produces JH until the fifth larval instar, and then it stops. The absence of the JH message then causes the fifth instar to develop into an adult insect. As long as JH is present, each larval molt leads to a larger larval form. When JH is absent, a larva develops into an adult at its next molt.

We can now explain the mystery of Sláma's linden bugs. The paper factor has the same effect on larval linden bugs as JH does. Linden bugs growing on a paper towel absorb juvabione into their bodies, and the protein receptor for JH responds to a molecule of juvabione just as it does to JH. The larvae stop making JH when they reach the fifth instar, but they continue to absorb juvabione. In a serious case of molecular mistaken identity, the receptor "recognizes" juvabione as JH and behaves as though it has received the JH message, "Do not develop into an adult." Acting on this message, the bugs continue to grow and molt, but they remain immature larvae.

Juvabione was one of the early examples of a plant chemical with JH-activity. Biologists later discovered that compounds with JH-activity are widespread in conifers and flowering plants. Although natural JH consists of only three or four closely related chemicals and varies little from one insect species to another, these JH-mimics include a wide variety of chemical types, many of which have chemical structures quite different from the natural hormone. Apparently, the protein receptor for JH is not very selective but accepts molecules with rather diverse shapes. Also, some of the JH-mimics are active in one group of insects but not others, indicating that the receptor is not the same in different groups of insects.

JH-mimics disrupt vital physiological processes in insects and can kill them, so they appear to be useful to the plant as defensive substances. One of the strong pieces of evidence supporting this assumption is that some JH-mimics seem to be present only when an insect predator is attacking the plant. In other words, the plant makes these compounds in response to insect predation.

These mimics provide a useful defense only when an insect predator is not making its own JH, but insects typically have a high level of JH through most of their lives. It disappears only as they become adults, and the adults may be short-lived. A more efficient plant defense would be an anti-JH compound, something that prevented the larval insect from synthesizing JH when it was needed. We do not know why such "antihormones" are not widespread. They would effectively turn off the JH message in larvae, causing them to become adults prematurely. This would be a useful defense against insects that attack plants only as larvae. Entomologists succeeded surgically in turning off JH in larvae many years ago by removing the gland that produces the hormone. Half-grown larvae thus treated turn into miniature versions of the adult insect, in a process known as precocious metamorphosis. Recognizing the defensive potential of anti-JH compounds, William S. Bowers and his colleagues (then working at the New York State Agricultural Experiment Station in Geneva, New York) screened a variety of plants for anti-JH activity. They discovered two compounds called precocenes that caused precocious metamorphosis in several species of larval bugs. If Bowers administered both precocene and JH to the larvae, development progressed normally, indicating that the precocene suppressed the larva's own synthesis of JH.

The discovery of plant compounds that either mimic or block the action of JH has had significant practical consequences. Both JH-mimics and anti-JH compounds are attractive replacements for conventional insecticides, because they are not toxic. Following this "biorational strategy," industrial laboratories have developed commercial JH-mimics and anti-JH compounds that have been marketed successfully as "insect growth regulators." Because JH-mimics prevent the development of adult forms, they are useful in controlling insects that are pests as adults, such as mosquitoes, fleas, and dung-breeding flies. Anti-JH compounds, on the other hand, are effective in control of insects that are pests as larvae, because they wipe out the larvae through precocious metamorphosis. This second group includes most of the significant agricultural pests—such as assorted armyworms, bollworms, cutworms, and tent caterpillars—that strike grains, fruit trees, cotton, and other major crops. Unfortunately, practical application of anti-JH compounds has been slow to develop.

There is another insect hormone called ecdysone that provides messages crucial to molting and pupation. Ecdysone is chemically quite different from JH but related in chemical structure to cholesterol. About a year after German scientists first obtained ecdysone from insects and identified it chemically, a research group in Japan found a very similar compound in a plant and showed that it mimicked the action of ecdysone in insects. Realizing that ecdysone-mimics could disrupt insect development and provide a new line of chemical defense in plants, a number of chemists began searching for other plant compounds akin to the hormone. They screened over one hundred plant families over the next few years and discovered some seventy ecdysone-like compounds. Several of these compounds are quite active as molting hormones in insects, and often they are present in high concentration in the plant. In general, it is the more advanced families of plants that contain these compounds. It appears that plant ecdysones are effective defense compounds, but the whole subject still needs more attention. We do not yet know, for example, what role plant ecdysones may have in the plant's own physiology.

PROTECTING POTATOES

Potatoes (*Solanum tuberosum*) are so common that it takes some effort to appreciate their importance. The potato plant is extremely adaptable, easy to grow, and does not require rich soil. In addition, it furnishes a highly nutritious, fat-free food at low cost. The International Potato Center in Lima, Peru, is devoted to helping developing countries around the world increase food production, devise sustainable agricultural systems, and improve human welfare through collaborative programs of potato research and development. It handles grants with a value of over $20 million a year. Because the potato is one of the world's leading food crops, agricultural experts devote considerable energy to developing new varieties that are better adapted to particular growing conditions and that are more resistant to predators. In the past twenty-five years they have screened over 180 species of wild potato, looking for desirable characteristics that can be incorporated into cultivated varieties through hybridization.

One of the most promising of these wild relatives of the cultivated potato is a hardy species, *Solanum berthaultii*, from the eastern slopes of the Andes Mountains in Bolivia. This plant is resistant to some of the most destructive predators of cultivated potatoes, including the Colorado potato beetle (*Leptinotarsa decemlineata*), various aphids, and the fungus (*Phytophthora infestans*) that causes the deadly disease known as late blight. The potato beetle is regarded as the most destructive pest of potatoes in the northeastern United States and is an increasingly severe problem in Europe. Aphids are very small insects that live on plant juices and transmit a number of viral diseases while feeding, much as mosquitoes transmit malaria. Late blight is the rapidly spreading infestation of potatoes that caused the Irish famine of 1845–46. Owing to agricultural improvements, it is no longer the threat that it once was, yet the disease can still be a problem in certain locations. Because of the broad resistance of the Bolivian wild potato, this species has attracted the attention of agricultural scientists for years. Investigators at Cornell University have been working with it since 1977, both to create useful hybrids with the cultivated potato and to explore the basis of its resistance.

We now know that the secret of the Bolivian wild potato's hardiness lies in little hairs, or trichomes, that cover its leaves. There are two kinds of trichomes, and together they mediate a remarkable chemical defense system. The type A trichome is about 200 micrometers in length (an ordinary grain of salt is 200 to 500 micrometers in diameter), and supports a spherical gland on its end. The gland is sectioned into quarters vertically, so the whole structure looks like a short stem topped with a small orange divided into only four sections. When the gland is disturbed, it separates from the trichome and breaks open, releasing a viscous liquid. The type B trichome is three or four times longer than the type A, and it continuously exudes a clear liquid that collects at its tip as a naked droplet. The liquid is sticky, and anything that brushes against the droplet picks it up. Roughly equal num-

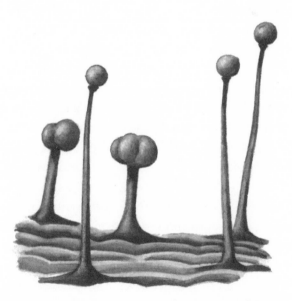

2.2 Tall type B trichomes and shorter type A's on a wild potato leaf. Cultivated potatoes never have type B trichomes, but they do very infrequently carry rudimentary type A's.

bers of the two sorts of trichomes are randomly distributed over the leaf surface about 100 micrometers apart. A photomicrograph of a leaf looks like an open forest of tall B's waving gracefully over short, stocky A's.

These trichomes carry a chemical defense system that any predator can set off. A very small predator such as an aphid (1 to 2 millimeters in length) is only a bit larger than the type B trichomes. As it moves across the leaf, it cannot avoid contact with the trichomes, and its legs soon get covered with sticky type B exudate. Moving becomes more difficult, and the insect stops frequently to groom itself. This exudate is largely a mixture of compounds related to ordinary table sugar (sucrose). These compounds are both adhesive and repugnant to the insect, and they inhibit its settling and feeding on the leaf. As the insect moves about, its sticky legs brush against the spherical glands of type A trichomes. The glands adhere to the legs. They become dislodged, burst open, and pour their viscous contents onto the insect.

This type A exudate contains chemical compounds called phenols as well as enzymes that specifically facilitate the oxidation and polymerization of these phenols. (Polymerization is the chemical combination of many small molecules to form a large molecule.) It is not yet clear just how chemical interaction of the phenols and enzymes is avoided in the gland before its rupture; perhaps the two reactive components are held in separate compartments. In any event, oxidation and polymerization, which also use oxygen from the air, begin only after the liquid has spilled onto the invading insect. Over the next several hours the type A liquid sets to a dark, hard mass, immobilizing the insect or severely restricting the movement of its legs and mouth parts. In addition, the exudates contain a mixture of volatile components that agitate the insect. One of the components is a common plant oil known as *E*-beta-farnesene, which also happens to be a chemical signal for several kinds of aphids. It delivers a message to the insects that says, "Flee! flee!" If the predator is an aphid that recognizes the signal, it will attempt to escape. Perhaps it will succeed, or perhaps it is so entrapped that it can only thrash about, breaking open more type A glands. These encounters frequently end in the death of the invader.

A larger predator, such as a potato beetle, is too powerful to become mired in these defenses. Its legs get covered with the exudates without impairing its mobility. However, the potato beetle does not like this wild potato. Adult female beetles are reluctant to feed or lay eggs on the leaves.

Many newly hatched beetle larvae will starve to death rather than eat the leaves. Something not yet identified in the type A exudate leads to this aversion, and its effects are enhanced by the sugar-like compounds in the type B exudate. Investigators have found that if they first remove the exudates by washing and wiping the leaves, potato beetles no longer reject the plant. Just how this resistance works on a chemical level is not yet clear.

Agricultural scientists are particularly interested in incorporating the chemical defenses of S. berthaultii into cultivated potatoes. One of their general problems in devising new means of plant protection is that, in time, insect predators evolve effective responses to plant defense systems. Because this Bolivian potato has relatively complex defenses that depend upon both chemical and mechanical effects, scientists believe that aphids, beetles, and other pests will be slow to evolve adequate responses. They have now created hybrid potatoes that incorporate functioning A and B trichomes. A remaining problem is that hybrids with type B trichomes have also acquired undesirable traits, such as reduced yields and late maturity, that are typical of the wild parent. Further selective breeding will be necessary to eliminate these undesirable traits from the cultivated hybrids. If these experiments are successful, the hybrid potatoes could be commercially attractive. Field tests have already indicated that their increased resistance translates into at least a 40-percent reduction in needed insecticides.

A DESERT CHEMICAL FACTORY

One of the Bolivian wild potato's defenses takes advantage of natural phenols and enzymes known as phenoloxidases to immobilize predators. Other plants contain related phenols and enzymes but use them for different purposes. Many plant scientists believe that these chemicals decrease available nutrients through a multistage process that appears to work in the following way. The living plant contains both phenols and phenoloxidases in its tissues, stored in separate compartments, so that they do not interact chemically. When a herbivore chews a bit of leaf or stem, the compartmentalization breaks down, and the phenols and enzymes come together. As the bite passes into the herbivore's gut, the plant enzymes facilitate reaction of the phenols with oxygen in the gut, and the phenols are converted to new compounds called quinones.

These quinones are chemically quite reactive, and they attack the plant proteins that were also in the bite taken by the herbivore. This reaction with the quinones modifies the proteins so that they are no longer readily digested by the herbivore. In addition, the quinones attack the herbivore's digestive enzymes, which are also proteins, and reduce their effectiveness in digesting the bite of plant. Now, the purpose of the herbivore's feeding on the plant in the first place was to get nourishment, and a large part of this nourishment would come from digesting plant proteins. The phenol-enzyme system has modified the proteins and inhibited such digestion, so the predator receives little nourishment from feeding on the plant, even though it has a full gut. This type of plant chemical warfare has the effect of starving the predator.

One of the most remarkable phenol-producing plants that apparently works in this way is the creosote bush (*Larrea tridentata*), a dominant desert shrub of the southwestern United States and northern Mexico. The creosote bush synthesizes several hundred volatile compounds and takes its name from the strong odor these chemicals impart. It is not necessary to crush a leaf to release the odor; it is unmistakable at a distance of several feet, and a hot desert breeze blowing across a cluster of creosote bushes smells vaguely like a chemical laboratory. The bush also exudes an amber resin that solidifies and collects on the leaves and stems, making up about 18 percent of the dry weight of the leaves. This resin is rich in phenols, and its unpleasant taste probably discourages foraging by mammalian herbivores. The leaves themselves contain phenoloxidases, so that on chewing a resin-covered leaf, a predator mixes the enzymes and phenols. As we have said, this combination is thought to lead to reactive, protein-destroying quinones that discourage predation. Many insects and other herbivores leave the creosote bush alone, presumably in deference to its potent chemical defenses. Most of those that do forage on it prefer mature leaves rather than young ones. Although mature leaves are tougher, they contain less resin, and this seems to be what matters.

The creosote bush fights off other plants as well as herbivores. Few other plants grow near creosote, probably because some of its chemicals are noxious to them. In the desert, where water is a limiting resource, the creosote bush keeps other plants away and thus reduces the competition for moisture.

The creosote bush has yet another method of laying claim to a large territory: it occupies more and more space by relentless asexual reproduction. A mature plant sends up new shoots from the roots at its outer periphery. These shoots grow into new plants, so that an irregular ring of bushes slowly grows up around the original one. As these new plants mature, they too send up shoots at their periphery, extending the circle further. As the older plants at the center die, they leave behind dead roots and an empty space in an expanding clump of creosote bushes, all of which are clones (genetically identical copies) of the original plant. Creosote bushes grow slowly, and these clumps can expand for hundreds of years. One irregular clump discovered near Old Woman Springs, California, was 22 meters across at its longest dimension. Radiocarbon dating and annual growth rings indicated that this huge clone of creosote bushes was well over nine thousand years old!

Creosote has a close relative, *Larrea cuneifolia*, that is native to the Monte desert of southern Argentina and has a defense system comparable to that of the North American species of *Larrea*. The South American desert grasshopper (*Astroma quadrilobatum*), however, feeds on this plant by choice. It is not unusual to find as many as one hundred adult grasshoppers feeding enthusiastically on a single medium-sized bush. These insects can strip the bush bare, greedily consuming leaves, flowers, and flower buds down to the woody stems. They seem to ignore the plants' defenses.

How does the grasshopper thwart the plant's chemical defense system? Biologists who have studied the problem offer a surprising explanation. This desert insect has in its gut its own phenoloxidase that competes with the enzymes from the plant and works faster than they do. The grasshopper has only a limited amount of oxygen available to oxidize the plant phenols. The grasshopper enzyme quickly depletes this oxygen in facilitating its own reactions of the plant phenols. However, the reactions facilitated by the grasshopper enzyme do not lead to reactive quinones, but instead to unreactive products that do not affect the plant proteins. Thus, when the slower acting enzymes from the plant begin to work in the gut, oxidation of the phenols to reactive quinones cannot proceed far, because all the available oxygen is used up before much quinone is formed. The plant proteins are left undisturbed, and the grasshopper digests them with no difficulty. For this South American grasshopper, a bush rich in chemical defenses is a nourishing food source upon which it can feed nearly uncontested.

IS THE LIMA BEAN'S SIGNAL A CRY FOR HELP?

Chemicals deployed by plants can poison a hopeful feeder, or interfere with its normal physiology, or drive it away. Some starve a predator by entrapment, or reduce the nourishment it derives from feeding. In each case, the plant launches a chemical attack on its enemy. In addition to direct counterattacks on marauding herbivores, plants under assault sometimes broadcast a chemical signal to bring assistance from a third species that preys on their herbivore enemy. In this kind of chemical warfare, chemicals function as signals rather than as weapons. This is a remarkable approach to self-defense that has come to light only recently. It is not yet completely understood, but in at least two instances we have enough information to suggest how it works.

One example of this three-species defense system involves the lima bean plant and two kinds of mites. The lima bean (*Phaseolus lunatus*) is a common agricultural plant, and mites are minute arachnids, eight-legged arthropods whose closest relatives are ticks and spiders. (The arthropods are a vast group of animals that also includes insects and crustaceans, such as crabs.) There are many thousands of kinds of mites, and they have parasitized virtually every known plant and animal larger than themselves. A group of these creatures known as spider mites are familiar to gardeners, and one particularly prevalent spider mite is the two-spotted *Tetranychus urticae*, which feeds on many agricultural and horticultural plants, including the lima bean. The other mite of interest regarding the lima bean is *Phytoseiulus persimilis*, a carnivorous mite that feeds on two-spotted spider mites and can exterminate whole populations of them. For convenience, we will call the two creatures simply spider mites and carnivorous mites.

Spider mites live by injecting saliva into plant leaves to dissolve the tissues and then sucking up the contents of leaf cells. They are intemperate feeders, and in the absence of predators, they will overexploit a food source, destroying it and consequently destroying large numbers of themselves. Always seeking a meal, spider mites are attracted to the leaves of a healthy lima bean plant by the mixture of volatile chemicals it emits. The mites settle on the leaves and commence feeding. Somehow, the mites' activity causes the bean plant to alter the mix of volatile chemicals that it releases. Those leaves besieged by the mites, as well as those untouched, begin to

28

send out a "distress signal." This new signal has a slightly different odor from that of an unstressed plant, and it carries several messages, each for a different recipient. It has a message for nearby lima bean plants that are not under mite attack. When the signal reaches them, these plants apparently also begin to send out the distress signal, even though they themselves are free of mites. The distress signal also reaches distant spider mites, but instead of attracting these mites, as the odor of an unstressed bean plant did, this new odor now repels them. They no longer come to forage on the plant. Finally, the third recipient of the new signal is the carnivorous mites, which are recruited to devour the spider mites.

The distress signal does not actually attract the carnivorous mites, because the mites cannot control their travels the way a bee or butterfly can. They are so small that when airborne they are at the mercy of wind currents as they wander the world in search of food. They reach the lima bean plant on a breeze only by chance, but once they land on a leaf, the bean plant's distress signal causes them to stay. Instead of taking off again on the next puff of wind, the carnivorous mites remain on the leaf as long as they receive the signal.

Interestingly, it is not the odor of the spider mites that attracts or retains these carnivorous mites. The critical signal comes from the lima bean plant. If the carnivorous mites land on a plant under siege, they find themselves a meal of spider mites. If they land on a healthy plant near the site of a spider mite siege, the carnivorous mites offer protection against spreading of the infestation. Either way, the lima bean plant has successfully enlisted the help of the carnivorous mites in defending itself from the ravenous spider mites.

One compelling question about this three-species communication system is who controls the signals. The Dutch agricultural scientist Marcel Dicke, whose doctoral thesis dealt with the lima bean system, has pointed out two possibilities. In the first, the spider mites land on the bean leaf, start feeding, and quickly take control of the plant's signal system. Under their direction, the bean plant ceases to attract more spider mites but rather sends the altered signal that repels their arrival. Mites already feeding on the plant effectively tell distant spider mites to stay away from this plant and to seek food elsewhere. This behavior lowers competition among the mites for their food resources. In this interpretation, the local spider mites direct the plant to signal distant spider mites. The carnivorous mites have simply learned to take advantage of the spider mites' communication to locate their own prey. Having lost control of its signal system, the lima bean plant is merely a passive beneficiary of the carnivorous mites' eavesdropping.

In Dicke's second interpretation, when the spider mites attack and begin to feed, the lima bean plant actively recruits carnivorous mites to its rescue with its distress signal. It is the distant spider mites that eavesdrop. They intercept the plant distress signal and understand it as marking a plant to avoid. Dicke favors the second explanation, in which the lima bean broadcasts the signal primarily for its own defense. He argues that a system primarily for communication among spider mites would have evolved along safer and simpler lines, avoiding volatile chemicals and the associated risk of attracting the spider mites' enemies. A choice between these two interpretations of the lima bean signal, or some third explanation, will be possible only when we know more about this three-species system.

Other plant scientists, working at the U.S. Department of Agriculture (USDA) Laboratory in Gainesville, Florida, are investigating a different three-species system that involves one of our most important crop plants. They have found that a seedling corn plant (*Zea mays*) under attack by a caterpillar calls on a tiny wasp (*Cotesia marginiventris*) for help. The attacking caterpillar is a moth larva known as the beet armyworm (*Spodoptera exigua*), and it feeds on the foliage of corn, sugar beets, and other crops. When beet armyworms chew on the leaf of a corn seedling, the plant picks up a chemical in "caterpillar spit," the liquid from the armyworms' mouths. The chemical acts as a signal to the corn plant, which then releases a mixture of volatile compounds to attract female wasps that are ready to lay their eggs.

These wasps, like many related species, parasitize the larvae of other insects and lay their eggs in them. *C. marginiventris* lays its eggs in several different caterpillars that are significant agricultural pests, including the corn earworm (*Helicoverpa zea*) and the fall armyworm (*Spodoptera frugiperda*), as well as the beet armyworm. The wasp eggs develop in the caterpillars' bodies. When they hatch, the wasp larvae use the caterpillars as both home and food, eventually killing the caterpillars. Parasitic wasps can cause high mortality in the species they strike, and they are important natural agents in controlling pest populations.

So far, it looks like the corn plants are rallying the egg-laying wasps to their defense. The USDA scientists have carried out experiments, however, that suggest that the plants' emission of volatile compounds is not specifically a call for help. The agricultural scientists think that these compounds are primarily feeding deterrents, since armyworms prefer to feed on leaves that are not producing volatiles. Also, corn plants release the same mixture of volatiles in response to "grasshopper spit," although wasps that come in response can do nothing to defend the plant from grasshoppers.

Another observation that suggests the corn plants are not really calling the wasps is the importance of learning in the wasps' behavior. The female wasps learn from experience. They respond to whatever odors they associate with the armyworms, using odor cues to help locate sites for their eggs. Both armyworms and their feces are essentially odorless, but a wasp that finds an armyworm associates the local odor with its find. If the odor is that of the corn plant volatiles, the wasp will be attracted to these volatiles the next time she is looking for suitable places to lay eggs. In the laboratory, the agricultural scientists associated armyworms with the odor of chocolate or strawberry. Although wasps rarely encounter these odors in a corn field, wasps that found armyworms in association with chocolate were attracted to chocolate in the future.

The natural three-species systems of corn and lima bean plants are reminiscent of an ancient strategy for pest control. Agriculturalists learned long ago that they could take advantage of predatory insects to control crop pests. A Chinese manuscript dated to A.D. 304 stresses the importance of introducing citrus ants into mandarin orange groves to keep the trees free of predatory insects. In our own era, plant scientists have repeatedly introduced predatory species into specific areas to combat destructive pests. In the natural three-species systems, predatory insects safeguard plants, but the plant itself, not the plant scientist, brings the rescuing predator.

Perhaps in the future we will be able to exploit natural three-species systems of plant protection in agriculture. These natural systems may offer practical advantages over applying broad-spectrum insecticides. Could we, for example, create a lima bean plant more resistant to spider mites than the commercial varieties now available? Different varieties of lima bean emit distress signals of different strength, so perhaps plant breeders could develop a variety with a much stronger signal than any now under cultivation. This new lima bean would recruit carnivorous mites more vigorously and be less susceptible to spider mites. This more proactive approach to pest control is a relatively new idea that has not yet been tested commercially with lima beans or any other crop plant. Other three-species systems will probably come to light now that we are aware of their existence, and they may operate in still other ways.

BUYING DEFENSE WITH A CHEMICAL BRIBE

With corn and lima bean plants we extended the weaponry of chemical warfare beyond noxious compounds to include chemicals that bring help from other species. Now we want to extend chemical warfare further to include compounds used by plants to buy protection from their enemies. A striking example of this chemical bribery comes from a group of trees called swollen-thorn acacias that "pay" certain ants to protect them from their enemies. The trees that have received the most intensive study (*Acacia cornigera* and related species) are native to Mexico and Central America, where they live in close and complex interaction with several species of acacia ants (*Pseudomyrmex ferruginea* and related species).

The trees actually offer more than a chemical bribe for the ants' services. They have developed some important structures not found in other species of acacias, and these structures provide the ants with both food and shelter. In swollen-thorn acacias, the bases of the thorns have become enlarged hollow structures that serve as nesting sites for the ants. The ants resident in one tree make up a single colony that may number up to thirty thousand individuals and occupy all the thorns of the tree. The trees also have special glands that secrete a sweet nectar directly onto their leaf stems. A 2-meter tree can produce as much as 1 milliliter of nectar a day, and this thick syrup, rich in sugars, is an important food for the acacia ants. They gather it with great care, and it provides essentially all the carbohydrate in their diet.

The nourishing sugary nectar is the chemical payment that the tree offers for the ants' protective services, but it is only one of the benefits the ants derive from associating with the tree. Virtually all the rest of the ants' food also comes from their acacia tree. They feed almost exclusively on special small leaves that are rich in protein and fat. Swollen-thorn acacias produce these leaves throughout the year, and so assure the ants of a continuing food supply. Other *Acacia* species do not have these special leaves and typically do not keep their leaves year round.

In return for food and shelter, the resident acacia ants guard their tree and keep it free of herbivores, treating the tree as an valuable resource. They attack any other insects they find on the tree, biting and stinging and driving them away. They kill very small insects and sometimes feed them to their larvae. They also sting and annoy foraging mammals, and can induce even large creatures to look elsewhere for food. Acacia ants are extremely aggressive and fast moving, locating intruders by both sight and contact. To humans, an acacia ant's sting is much more painful than that of closely related species. In their zeal to maintain their acacia tree, the ants even eliminate the leaves of foreign plants that are close enough to touch the tree's foliage or that grow beneath it within a radius of a meter or more. These defensive activities keep the acacia free of herbivores and furnish it with a sunny, open growing space. This is important, because these trees do not tolerate shade well.

The interaction of ants and acacias clearly confers advantages on both organisms. Moreover, these advantages are not merely desirable consequences; they are absolute necessities. Daniel H. Janzen, then working at the University of Kansas, found that neither acacia ants nor swollen-thorn acacias can survive outside their interdependent relationship. Acacia ants are unable to adapt to life without the shelter, nectar, and leaves that the acacias provide. Without the trees, they die. For their part, swollen-thorn acacias lack the effective defenses they would need to live on their own. Unlike acacias that do not support ant colonies, their leaves contain no bitter substances to deter foraging herbivores. Without acacia ants to patrol their branches, the trees become attractive targets for predaceous insects. Their nectar glands lure beetles, flies, and cockroaches to imbibe the readily accessible nourishment. The predators then fan out over the trees to feed on the foliage. Deprived of its ant colony, a swollen-thorn acacia usually succumbs within a year, but a tree with a thriving ant colony may live for twenty years.

*W*E BEGAN our discussion with the bitter cucurbiticins of gourds and squashes and considered several kinds of defensive chemicals, finally reaching the sugars of swollen-thorn acacias. Perhaps we stretched our definition of chemical defense a bit to consider common sugars as agents of chemical warfare, but doing so reveals the many ways that plants protect themselves chemically.

CHAPTER

3

HOW ANIMALS USE
CHEMICAL WARFARE

*C*HEMICAL DEFENSES in the animal world are perhaps better known than those used by plants. We are all familiar with animals that have poisonous stings and bites, and many of us have had a painful experience with a hostile bee or wasp. Aggressive so-called killer bees have arrived in the United States from the south, and serious bee stings are in the news along the Mexican border. Stinging animals have been familiar nuisances for millennia. Aristotle (384–322 B.C.) mentions several different species of them in his *History of Animals* and reports that "a horse has been known to have been stung to death by [bees]." Much earlier, thousands of years before the Greeks, cave painters depicted men robbing honey from wild bees. They must have been very familiar with bee stings.

Painful stings and other poisons can provide animals with an effective defense, but they are only one of many kinds of chemicals that animals employ to protect themselves or combat their enemies. Skunks repel attackers with a malodorous spray. Soldier termites discharge chemicals that entangle their insect enemies in a sticky mass. Slave-making ants disperse chemicals that spread terror and dissension among their victims during slave raids on other ant colonies. These are all techniques of aggressive chemical warfare, and we could reasonably limit our attention to such activities. We shall once again take a broader view, however, and also consider some less aggressive ways that animals protect themselves with chemicals.

This broader view permits us to regard as chemical defenses the deceptive chemical signals used by certain beetles to sustain their parasitic way of life. These parasitic beetles make their homes in the nests of ants or termites. They live off the bounty stored up by their hosts, and sometimes they feed on host larvae and workers. If the hosts discovered these intruders, they would quickly eject them from their nest, but the beetles employ a clever ruse to avoid detection. They carry on their outer surface, or cuticle, the same chemical compounds that are found on their hosts' cuticle. When ant or termite nestmates encounter each other, each one touches its antennae to the other's cuticle. Biologists believe that the insects are verifying each other's cuticular compounds and that these compounds are a recognition signal identifying members of the same nest. The beetles offer the proper recognition signal to their hosts and so pass themselves off as members of the colony. Recognition chemicals provide the parasitic beetles an indispensable defense from detection by their unwitting hosts.

Whereas these beetles simply carry their camouflaging compounds on their outer surface, other animals have different ways of deploying their defenses. Bees and snakes actively inject venom into their adversaries, while several butterflies and their caterpillars passively carry protective toxins throughout their bodies. (Poisons of biological origin are usually referred to as toxins, and a toxic secretion that is transferred to a victim by biting or stinging is known as a venom.) Grasshoppers smear their liquid repellent on attackers, and skunks spray theirs. Some soldier termites' heads are equipped with brushes or nozzles for wiping or squirting noxious agents on nest invaders. Other termites resort to dramatic self-destruction. With a sharp contraction of their abdomen, they explode, covering invaders with a viscous mass of internal organs and blood. This is surely the ultimate chemical defense, as these insects turn themselves into chemical bombs.

There are also animals that spread their defensive compounds onto objects rather than other organisms. The tiny Brazilian wasp, *Mischocyttarus drewseni*, builds a small nest that hangs from a stem several inches long. The wasps' archenemies are ants, and the only way for ants to invade the wasp colony is to crawl along this nest stem to the nest. The wasps protect their nest from ant attack by wiping a repellent secretion from their abdomens

3.1 This termite soldier of the species Nasutitermes corniger *squirts a gluey defensive liquid at enemies through its long snout. Termites are blind, and the soldiers aim their fire in response to air currents caused by the enemy's movement.*

along the stem. Ants searching for food discover the stem and begin to crawl along it toward the wasp nest, but on encountering the secretion, they turn back and leave the wasps in peace.

Chemical defenses are much more important for some groups of animals than for others, although they are found throughout the animal world. Among the vertebrates, fishes make great use of chemical defense, but mammals and birds much less so. Until recently, scientists knew that the flesh or eggs of some birds tasted bad and that predators refused them as food, but no one had obtained any unpalatable or toxic chemical from a bird. In November 1992, however, a team of biologists working in New Guinea found a very poisonous compound in the skin and feathers of a bird known as the hooded pitohui (*Pitohui dichrous*). The local people were already aware of the pitohui's toxicity. They eat the bird only after its skin is removed and the meat is then prepared in a special manner.

The pitohui's toxic compound is chemically complicated, and had turned up earlier in three species of frogs that live in the forests of Colombia. Perhaps all these creatures have independently learned to synthesize the same unusual compound, or perhaps they all get it from a similar but undiscovered external source, such as a microorganism or parasite common to them all. Neither explanation is very satisfying in accounting for a toxin that has turned up only in a bird from the South Pacific and in three frogs half a world away.

As with fishes, many groups of insects have chemical defenses. In general, organisms with a high probability of being discovered by other species are more likely to have chemical defenses than those that lead hidden lives, and organisms that have well-developed alternative means of defense are less likely to have chemical defenses than those without other protection.

Defensive chemicals sometimes serve an animal in other ways as well. A toxin can be a good defense and also an important offensive weapon. No one bitten by a rattlesnake doubts its potent means of deterring enemies, but rattlesnakes also routinely bite and inject venom into prey that they themselves seek out. Venom is an important tool in gathering their food. Similarly, many wasps that live in colonies sting intruders that threaten their nest, and they also sting their prey. They spend their days away from the nest hunting caterpillars that they kill with a well-placed sting. They butcher the caterpillars on the spot and carry the pieces home to feed their larvae.

THE BOMBARDIER BEETLE'S MUNITIONS FACTORY

Some animals under assault counterattack with an irritating spray that causes local pain or discomfort. Such a spray may even inundate small insect predators. Vertebrate predators are too large to be inundated, but they are particularly sensitive to chemicals about the eyes and mouth. Whatever their size, predators often break off an attack to rid themselves of an irritant. A mouse cleans its eyes and snout with its feet or sometimes rubs its muzzle into the ground. A bird wipes its head repeatedly against its feathers. An ant rolls in the dirt to wipe itself clean. While the irritant distracts the predator, the intended victim flees to safety.

There are many kinds of chemical irritants, and animals spit, spray, or squirt them at their predators. One amazing delivery system belongs to about thirty species of insects known as bombardier beetles (*Brachinus* species). These are ground beetles, 8 to 12 millimeters long, and are common insects in North America and Europe. Most bombardier beetles spend the daylight hours under rocks or logs. They are active at night and feed on the larvae of both moths and other kinds of beetles, many of which are economically significant pests. These characteristics are unexceptional; many other ground beetles lead similar lives. Only their chemical defense sets bombardier beetles apart.

When a bombardier beetle is attacked, it responds with a spurt of liquid accompanied by an audible pop. The liquid is quite hot and contains a mixture of irritant quinones. The spurt is aimed directly at the aggressor. The explosive pop, which gives the beetle its name, is loud enough to startle the aggressor, and the liquid itself burns and distracts it. All this drama has made bombardiers a favorite of naturalists and entomologists since the eighteenth century. The most detailed investigations of these insects are those that Thomas Eisner and Hermann Schildknecht independently carried out over thirty years ago. Eisner is a biologist whom we shall meet again shortly, and Schildknecht is a chemist who was then working at the University of Erlangen in Germany.

Eisner studied the effectiveness of the beetle's defense system. He gathered bombardier beetles for his investigations from beneath rocks near a local swamp and maintained them in the laboratory on a diet of insect larvae. Each beetle lived in its own small glass dish. To carry out his experiments, Eisner needed a way to manipulate the beetles without triggering their defensive reaction. He did this by attaching a very small aluminum hook to the back of each insect with a bit of liquid cement. He could then grasp the hook and move the beetle about without incident. By connecting the hook to a clamped rod, he could also immobilize the beetle when he wished.

Eisner wanted to learn how accurately the beetles could aim their spray. To find out, he recorded the path of the spray using common laboratory indicator paper. By placing an immobilized beetle on a small piece of the paper and then provoking it, Eisner could obtain a permanent record of the beetle's defensive shot. When provoked, the beetle responded with a rapid pop and a spurt of quinones. The quinones turned the white indicator paper an intense blue-black color wherever the spray struck, and this provided a visible record of the shot.

With this setup, Eisner discovered that the beetle is an accurate marksman and can direct its irritant precisely to the site of attack. Eisner provoked his immobilized beetle by pinching one leg with jeweller's forceps or touching its back with a hot needle. Each time it was attacked, the beetle

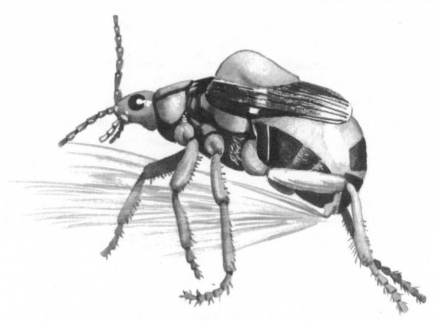

3.2 Pinching a bombardier beetle's leg elicits a well-aimed spurt of hot chemicals. An accurate description of this distinctive chemical defense appeared as early as 1750.

replied with a shot right to the annoying forceps or needle. Eisner found that the beetle's accurate aim is possible because the end of its abdomen is very flexible in all directions. The insect can move the tip around and aim its discharge wherever it wishes. Some beetles could fire more than twenty times in four minutes before exhausting their supply of chemicals.

Schildknecht carried out chemical experiments that revealed how this defense system works. Each time a beetle fires a shot, it performs chemical reactions in a pair of glands located at the rear of its abdomen. Each gland consists of two chambers. One chamber is a reservoir that holds hydrogen peroxide and a mixture of phenols. The other chamber contains a mixture of common enzymes known as catalases and peroxidases. This second chamber has heavy walls and is the beetle's chemical reactor. The two chambers are connected by a channel with a valve that the beetle can open or close. A second channel leads from the heavy-walled reactor directly to an opening at the flexible tip of the insect's abdomen.

When the beetle is disturbed, it opens the valve between the two chambers. Hydrogen peroxide and phenols pass immediately from the reservoir into the reactor. There they mix with the enzymes already present. The enzymes catalyze the rapid formation of oxygen from hydrogen peroxide, and the phenols are then rapidly oxidized to quinones. Oxidation of the phenols is a sudden, explosive reaction that releases a significant amount of heat. The temperature of the liquid in the reactor quickly rises to the boiling point of water, and the explosion forces the hot liquid out through the tip of the abdomen with a popping noise. The beetle has already aimed the tip properly, and the shot is on its way toward the attacker.

In this one set of chemical reactions a bombardier beetle synthesizes a solution of defensive quinones, heats the solution to the temperature of boiling water, generates sufficient force to fire the solution at an assailant, and produces a surprising pop—impressive chemical warfare for a little beetle.

CHEMICALS SYNTHESIZED AND BORROWED

Chemical defensive agents serve many purposes, but animals ordinarily acquire them in one of two ways. The first is for animals to synthesize their own compounds, just as plants synthesize the defensive chemicals they use. Animals that do this possess the metabolic machinery to convert chemicals in their diet into compounds for defense or offense. They make their chemicals for warfare the same way they make hormones, enzymes, and other chemicals that all creatures require.

The second way animals get their chemicals for warfare is simply to appropriate compounds from another species, often by feeding on the donating species and then sequestering the needed compounds. In either case, an organism must have the means to protect itself from the compounds' effects. Skunks fastidiously avoid contaminating themselves with their own spray. Butterflies that carry toxins are resistant to the effects of these poisons. On the other hand, bees are not immune to their own venom but protect themselves by storing it in a chamber isolated from the rest of their body. Two queen honey bees fighting for a colony sting each other repeatedly, until one of them is killed.

Gourd-eating beetles accumulate gourd toxins for their own future

41

protection. Some creatures with a more diverse diet can accumulate and use many different chemicals. One animal that does this is the lubber grasshopper, *Romalea guttata*. Lubbers are large, slow-moving, flightless insects native to the southeastern United States. They feed on green leaves from many different plants, as well as on fruit, bark, fungi, and dead plant and insect matter. Like many other grasshoppers, lubbers regurgitate a dark brown liquid when they are disturbed. This grasshopper spit contains as many as thirty or forty defensive compounds. It is a mixture repugnant to many predators, who find lubbers unpalatable. Owing to lubbers' highly varied diet, the regurgitated mixture differs from one individual to another. The quality, quantity, and effectiveness of each grasshopper's chemical defense depend on what it has eaten recently. Many species of animals have the ability to accumulate plant defensive compounds from their diet as lubbers do, but the details of how they do this are not yet clear. How does an animal choose which compounds to store? How does it avoid simply metabolizing these compounds along with its other food? The questions are easy to ask, but there are no good answers yet.

Ingesting and storing a plant's defensive chemicals is not the only way to profit from them. Another possibility is to appropriate the plant itself. Birds that reoccupy old nests often do this very effectively. Many birds do not build new nests each time they breed, but instead refurbish nests they built earlier. This practice increases the likelihood that their nests will harbor parasites and microorganisms that can infest the adult birds and their young, since the nest has had the preceding season to accumulate such invaders.

The common starling (*Sturnus vulgaris*) is a bird that recycles its old nests, and it has a practical way of contending with the threat of foreign organisms. A small percentage of the material that starlings collect to refurbish old nests is fresh green vegetation. They select this material by odor, preferring the leaves and shoots of plants that are rich in volatile, odoriferous oils. These oils are plant defensive chemicals that are toxic or repellent to insects and microorganisms. The fresh vegetation successfully fumigates the old nests as the starlings rebuild them. The birds continue to freshen their nests, incorporating new bits of green vegetation into the reoccupied nests until their eggs hatch.

MURDEROUS DECEPTION AMONG THE FIREFLIES

Most animals that borrow poisons take them from plants, but one group of fireflies has a remarkable technique for obtaining toxins from other fireflies. Our knowledge of this unusual behavior comes from studies at Cornell University by biologist Thomas Eisner and chemist Jerrold Meinwald. These investigators and their research groups are responsible for many significant advances in chemical ecology over the past twenty-five years, particularly in the chemical defenses of insects. Eisner and Meinwald found that several species of *Photinus* fireflies carry a mixture of potent steroidal toxins that they named lucibufagins. Owing to the toxins, insectivorous birds and spiders refuse to eat these fireflies.

The Cornell investigators also found that the females of another genus of fireflies called *Photuris* also contain lucibufagins. Unlike the *Photinus* fireflies, however, the *Photuris* females do not make the toxins. They get them from eating *Photinus* males. Several species of *Photuris* are predators, and several species of *Photinus* are the victims upon which they prey. When the predatory females first emerge from their cocoons, they are not toxic, and spiders eagerly consume them. These same spiders refuse older females that have fed on *Photinus* males. Lucibufagins are a significant defense for the predatory *Photuris* females, but they can keep their level of the toxins high only by feeding regularly on their male prey.

The female predators need a constant supply of victims, and they catch them by employing an exceptionally clever trick. They have learned how their male prey search out females for mating and how they can intervene and take advantage of this search. The male *Photinus* fireflies use their lights to locate receptive females of their own species. In each prey species, the males have a unique flashing code that attracts a response only from females of the same species. In one species the males may flash, wait two seconds, and flash again, while males of a related species give perhaps five very short flashes evenly spaced over three seconds, and a third species sends three quick blips in half a second.

At night the *Photinus* males fly about flashing their signals to attract the attention of receptive females. The female fireflies are usually at rest on low vegetation or on the ground. When a female sees the right male signal for her species, she flashes the appropriate female response, which also differs from species to species. The male recognizes her response and flies toward her, flashing his signal again. In this way with repeated signals back and forth, he locates a prospective mate, flies to her, and lands to consummate their relationship.

This system is quite successful. Receptive females are mated within six minutes after they emerge from their underground burrows to look for a male. The females mate only once, but the males seek to mate repeatedly. There are roughly equal numbers of males and females, so that many males are flying and flashing in a continuing search for scarce receptive females. Unfortunately for the *Photinus* males, the predatory *Photuris* females have mastered the *Photinus* signalling system and exploit it ruthlessly for their own purposes. Our understanding of this ingenious behavior comes from extensive investigations by James E. Lloyd of the University of Florida.

When a male of a prey species flies and flashes, it is not only a female of his own species that may notice his light. A hungry predatory *Photuris* female may also be on watch. She knows the proper female response to the male signals of two or three *Photinus* species, and she returns the appropriate female signal to the flashing male. He sees this and assumes it comes from a receptive female of his own kind. He turns toward her and repeats his flash. She replies again, just as a receptive female should do. The male flies to his tryst, repeating his flash and suspecting nothing. But when he lands beside the predatory female, she simply takes him up and eats him. The female gets her meal and also her victim's toxins, which she stores for her own future protection against birds and other predators.

MILKWEED, MONARCH, AND FLY

Some animals that acquire chemical defenses from other animals are actually borrowing plant defenses at one remove. Certain herbivores sequester chemicals from plants and then pass them on to their own predators or parasites. This kind of extended sharing takes place among a milkweed plant, a caterpillar, and a parasitic fly. The milkweed (*Asclepias curassavica*) protects itself with a group of poisonous steroids known as cardenolides. (These are related chemically to the lucibufagins of *Photinus* fireflies.) The caterpillars of the monarch butterfly (*Danaus plexippus*) are insensitive to these toxins. They feed on milkweed leaves and retain the cardenolides. As a

44

result, most insectivorous birds refuse to eat monarch caterpillars or adult monarch butterflies. A parasitic fly (*Zenillia adamsoni*), however, is also resistant to cardenolide poisoning. When the female fly is ready to lay eggs, she seeks out a monarch caterpillar and carefully deposits her eggs on it. On hatching, the fly larvae bore into the caterpillar and consume it from within. They retain the caterpillar's cardenolides and maintain protective levels of these toxins in their own bodies.

Monarch caterpillars are not only filled with toxic cardenolides; they are also distinctively colored. They are ringed with black, white, and yellow bands that make them conspicuous as they feed on a green leaf. Many other poisonous animals are also colored in bright or striking patterns that are easily recognized. After one or two unpleasant experiences, predators associate the distinctive coloring of these animals with their toxicity and avoid attacking them. Instead of hiding or camouflaging themselves, these toxic creatures make sure that potential enemies recognize them and the danger they represent. Their safety lies in high visibility. Distinctive coloring is formed by chemical compounds, and without stretching our definition too far, we can regard coloration as another kind of chemical defense.

TWO WAYS TO THRIVE AS AN UNINVITED GUEST

Often it is not obvious whether an organism makes or borrows its chemical weapons, and animals with similar needs may acquire their compounds in dissimilar ways. Two species of parasitic beetles of the sort we mentioned earlier provide a good example of this. Both beetles live with social insects and deceive their hosts with cuticular recognition signals. One of the beetles, *Myrmecaphodius excavaticollis*, thrives in the nests of several species of *Solenopsis* ant. It fools its hosts by carrying on its surface the ants' cuticular compounds, which it acquires directly from association with the ants. Everything in the ants' nest is covered with these compounds, and beetles that invade the nest pick them up by contact.

Beetles maintained in the laboratory away from their hosts quickly lose these protective compounds and need a few days in the ant nest to regain them. A beetle is vulnerable for these first days in an ant nest without its protective recognition compounds. During this time, it tries to lay low and avoid being challenged by the ants. If the ants find and attack the beetle, its response is to play dead, and the ants are often deceived. Each species of

Solenopsis ant associated with this beetle carries a distinct mix of cuticular compounds, so the recognition signal that the beetle displays depends on which species is its host. The ability to pick up these compounds on contact gives the beetle flexibility in its living arrangements.

In contrast, *Trichopsenius frosti* is a beetle that synthesizes for itself the cuticular compounds of its host. *T. frosti* lives in termite nests, and its host termite, *Reticulitermes flavipes*, produces a cuticular mixture of at least twenty-one compounds. One component of the mixture accounts for over one third of the total, while some of the other components amount to well under 1 percent. The beetle produces a mixture containing all twenty-one components in amounts close to their abundance in the termites' mixture, a good enough imitation that the termites accept it without question. This beetle lives with only a single species of termite and has invested consider-able effort in duplicating the recognition signal that makes that lifestyle possible.

Each of these beetles has acquired the ability to deceive its host over a long period of development, change, and interaction with its host. In each case, the details are specific to the particular organisms and their association with each other. The result is two quite different solutions to very similar problems.

STEALING A UNIQUE WEAPONS SYSTEM

While most animals merely borrow a compound or two from other species, one group of marine organisms appropriates an entire chemical delivery system. The system itself is complicated and unique, and we should first say something about it and about the organisms that create it, before discussing the ones that steal it. This unusual system is the creation of a group of or-ganisms called coelenterates, or cnidarians, which comprise the most poi-sonous major group (phylum) of animals.

The cnidarians include some ten thousand species of jellyfish, sea anemones, corals, and their relatives. There are a few freshwater species, but nearly all of them live in the ocean. Not only are they all poisonous, but they all use the same general system to deliver their toxins to prey and ene-mies. Cnidarians package their toxins in tiny capsules called nematocysts. There are several types of nematocysts, and we shall focus here on one com-mon type. Within the nematocyst capsule there is a long hollow thread, or tubule, along with the toxin. The coiled tubule is attached from within to the top of the nematocyst, and the top is closed from the outside with a tight-fitting cover. Special cells containing these nematocysts are concentrated at the animal's outer surface, particularly along the tentacles that many cnidarians use to snare their prey.

Nematocyst cells are sensitive to touch and to the presence of sugars and other food chemicals in the water. When one of these cells is stimulated, it responds instantly with explosive force. Within only a few thousandths of a second, the nematocyst cover flies open, and the coiled tubule shoots out. As the tubule extends from the nematocyst, it turns inside out, just like the inverted finger of a rubber glove. The tubule is launched with so much force that it frequently penetrates the cnidarian's target. The base of the tubule remains attached to the top of the nematocyst, and the toxin inside the nematocyst is forced out through the tubule and into the target.

Most of the toxins delivered by this system are proteins, and their primary function is to stun or kill the cnidarian's prey. Some of these chemicals are among the most potent of all known biological poisons, so that a few species of cnidarians that live in waters frequented by human beings pose a public health hazard. Probably the best known of these dangerous species is the Portuguese man-of-war (*Physalia physalis*), which is common in warm oceans throughout the world, including the waters off the southeastern United States. Portuguese men-of-war ride on the ocean surface, supported by a vivid blue, balloonlike float several inches across. They carry their highly toxic venom in tentacles that radiate from the float and can be more than 60 feet long. Each year in the United States, several hundred swimmers are stung by Portuguese men-of-war when they become entangled in these tentacles. For persons with cardiovascular or immunological problems, the encounter can be life-threatening, and perhaps 10 percent of these stinging incidents are fatal. Less susceptible human victims often experience excruciating pain, difficulty in breathing, convulsions, and other unpleasant effects.

Cnidarians are simple animals that appeared on earth relatively early. Fossil species that lived 500 to 700 million years ago appear little different from some contemporary forms. It is striking then that cnidarian nematocysts are among the largest and most complex structures produced within the cells of any animal.

Modern biologists regard the production of nematocysts as one of the defining characteristics of the cnidarians. In the nineteenth century, however, scientists generally believed that other animals also generated nematocysts. They had discovered that cnidarian-eating mollusks called nudibranchs defend themselves by firing nematocysts at their enemies. While it seemed self-evident that these nudibranchs produced the nematocysts they used, a few biologists disagreed. They proposed that nudibranchs actually appropriate their nematocysts from the cnidarians they consume.

For decades no one took the idea that nudibranchs steal nematocysts seriously. Biologists all agreed that nudibranchs feed upon cnidarians, but most of them considered it very unlikely that intact cnidarian nematocysts could pass into the nudibranchs' cells and that nudibranchs could then fire these stolen nematocysts at their own enemies. Our understanding of nematocysts and nudibranchs changed only at the beginning of the twentieth century. Although it may seem unlikely, nudibranchs do indeed steal cnidarian nematocysts, in one of the most remarkable instances of thievery in nature.

Biologists now agree on the basic facts of this unusual story. The nudibranchs that interest us here are generally small, perhaps a few inches long. They prey on various cnidarians, such as sea anemones and corals. Nudibranchs are beautiful, soft-bodied, delicately colored mollusks. They have no shell, and their soft bodies are protected by fingerlike projections called cerata that extend from the upper surface of the animal. The interior of each of these cerata communicates directly with the nudibranch's central digestive cavity.

When a nudibranch feeds, its digestive cavity fills with bits of cnidarian tissue, including large numbers of nematocysts. Some of the nematocysts rupture during feeding, but their toxin does not disturb the nudibranch. Many nematocysts reach the nudibranch's digestive cavity intact. These nematcysts make their way from the central digestive cavity into the interior of the cerata. They move about, and some of them reach the outer tips of the cerata. Here there are cells that recognize, surround, and engulf the nematocysts. Each cell that takes up a nematocyst orients it in a particular outward direction and actively maintains it in good condition.

The nematocysts are now strategically positioned at the tips of the nudibranch's protective cerata. Cnidarian nematocysts that started out as primarily offensive weapons have now become the nudibranch's first line of defense. When it is attacked, the nudibranch ejects large numbers of nematocysts from its finger-like cerata, keeping the enemy away from the main portion of its body. The defense is effective, and many fishes and other predators decline to eat these armed nudibranchs.

These are the basic facts about nudibranchs and nematocysts, but many questions remain. How do nudibranchs avoid firing all these sensitive nematocysts as they incorporate them into their cerata? How do they avoid simply digesting the nematocysts along with the rest of the cnidarian tissue they ingest? How do the cells at the tips of the cerata recognize the nematocysts before they engulf them, and how do these cells maneuver the nematocysts into place and aim them so they can be fired beneficially to protect the nudibranch? Biologists are still exploring how nudibranchs solved these problems millions of years ago.

CHEMICAL DECEPTIONS AMONG THE MAMMALS

The most familiar chemical defense among the mammals must be the skunks' repellent spray. Skunk spray has a penetrating, sulfurous odor that no one can ignore or forget. Skunks (*Mephitis mephitis* and related species) appear at dusk and move fearlessly through woods and meadows, advertising their presence with distinctive black and white markings that remind everyone of their special threat. Predators apparently learn the lesson quickly, for few attack a skunk twice. Skunks are not particularly pugnacious and do not seek a fight. If overtaken or cornered, however, they stand their ground and fire. Automobiles fail to intimidate them, and many die when they wander onto highways. While other members of the weasel family also have anal scent glands they use for defense, the best-known repellents are those of the several species of New World skunks and Old World polecats.

Skunks and polecats are renowned for their chemicals, but biologists have discovered only a few other mammals that practice chemical warfare. One of the most ingenious defenses belongs to the hedgehogs, a family of small mammals related to moles and shrews, and native to Europe, Asia, and Africa. Hedgehogs are no longer found in the New World (they became extinct in North America about 10 million years ago), and they are known here primarily through Old World fairy tales and children's stories. In Britain and parts of continental Europe, one species of hedgehog, *Erinaceus europaeus*, is a common garden animal that is easily tamed and sometimes kept as a pet. Hedgehogs' backs are covered with stiff, sharp spines that provide them with an impressive structural defense. When threatened, they can raise the spines around their head and lunge at an attacker. More often, they roll into a tight ball so that their belly is protected and only the erect spines on their back are exposed. Hedgehogs can maintain this defensive posture for hours.

Many hedgehogs frequently display another peculiar behavior. They bite or chew on objects having a strong or unusual odor. As they chew, they begin to froth at the mouth, and then they wipe the frothy saliva over their spines with their tongues. This behavior is known as self-anointing. Some hedgehogs do it more readily than others, and some may even respond to a strong odor with self-anointing, even if there is no object to bite or chew on before frothing. One tame indoor hedgehog self-anointed when someone lighted a cigar in the room. Infant hedgehogs self-anoint several days before their eyes open.

Many biologists believe that this odd behavior derives from a chemical defense in which wild hedgehogs borrow toad toxins for their own protection. Toads live throughout the range of hedgehogs and are a common hedgehog food. When Edmund D. Brodie Jr., a biologist at Adelphi University, presented skinned dead toads to hedgehogs, the hedgehogs ate the toads without self-anointing. They also ate whole white mice and bullfrogs without self-anointing. When presented with live toads or unskinned dead toads of eight different species, however, the hedgehogs first chewed the toad's skin and self-anointed before feeding on the toad. Toad skin contains defensive toxins, and it appears that hedgehogs enhance the effectiveness of their spines by coating them with the toxins that they extract by chewing on the skin.

In Brodie's laboratory, human volunteers jabbed themselves with spines covered with secretions from toad skin and also with untreated spines. Not surprisingly, treated spines caused much more pain and inflammation than untreated ones. It seems likely that spines smeared with toxin-filled saliva protect hedgehogs better than untreated spines do.

Brodie's observations indicate that self-anointing with toad skin can make the spiny defenses of hedgehogs more effective. It is less clear how effective self-anointing may be when it is elicited by other, less toxic materials.

In addition, although some tame hedgehogs spend much of their time self-anointing, other tame ones do not self-anoint at all. Self-anointing appears to mean different things to different hedgehogs, and its general significance remains uncertain.

Regrettably, such uncertainty is not rare in interpreting mammalian behavior. Mammals lead complex lives and have large, intricate brains. These brains integrate information received from various sources, and mammalian responses are much less reproducible than those of more simply organized creatures. In interpreting the behavior of hedgehogs, there is the additional difficulty that many reports concern animals kept as pets. Tame animals live in a world quite unlike that of their wild relatives, and they often behave quite differently.

The complexities of mammalian behavior may be one reason why biologists know of so few mammalian chemical defenses. Perhaps there are mammals with chemical defenses that we simply do not recognize. The self-anointing of hedgehogs was familiar for many years before Brodie carried out his experiments and then proposed that self-anointing permits hedgehogs to appropriate toad toxins for their own defense.

There is another instance of self-anointing among mammals that seems clearly to be a borrowed chemical defense. When Siberian chipmunks (*Eutamias sibiricus asiaticus*) encounter a dead snake, they approach the carcass cautiously and first make certain that the snake is dead. Then they come up to the carcass and begin to bite and gnaw at the snake's skin. They chew on bits of skin and then twist around and apply them to their fur. In controlled field tests, chipmunks responded with this behavior to the carcasses of four different species of snakes, as well as to snake urine or feces. In contrast, they showed no interest in the carcasses, urine, or feces of various lizards, frogs, birds, and mammals.

The biologists who studied these chipmunks, Tomomichi Kobayashi and Munetaka Watanabe of Okayama University in Japan, also found that snakes are less likely to attack prey animals that are covered with bits of snake skin, urine, or feces. Because snakes prey on Siberian chipmunks, this behavior implies that self-anointing provides the chipmunks a meaningful deterrent from attack. It would be interesting to know whether the snake odor also deters foxes, badgers, and other animals that feed on Siberian chipmunks.

51

ANIMAL TOXINS BORROWED BY HUMANS

Like many other chemicals fashioned by evolution to serve specific purposes, animal toxins often have unique properties not yet duplicated by chemical compounds synthesized in the laboratory. We use them not for chemical warfare, but often as unique tools for biomedical research, especially for studying the transmission and reception of nerve impulses. Different types of nerves make use of different chemical compounds to pass signals on, and there are specific receptors for each of these compounds. Some animal toxins affect one or another of these receptors and thus interfere with a specific type of nerve impulse. By working with a particular toxin, an investigator can study the action of the receptors affected by that toxin. Neuroscientists understand one type of receptor associated with muscle cells, for example, better than all others largely because they have available a snake venom that facilitates research on this particular receptor. They also exploit scorpion toxins, spider venoms, and some other poisons as important adjuncts in their research.

One of the most interesting of these research tools is an unusual compound known as tetrodotoxin that is found in puffer fish (*Fugu vermicularis vermicularis* and several related species). Owing to its high efficiency in disrupting nerve impulses, tetrodotoxin is exceedingly poisonous. Studies in laboratory animals suggest that a dose weighing slightly less than 1 milligram (the weight of three or four ordinary grains of salt) can be lethal for humans. The toxicity of puffer fish has been familiar for thousands of years. It was apparently known to the ancient Egyptians, and it is mentioned both in *The Herbal*, a Chinese pharmacopeia written in the first or second century B.C., and in a Japanese work, *The Book of Medicine or Herbs*, that appeared about a thousand years later.

Because human beings learned so long ago that puffer fish are toxic, we might assume that they also learned to avoid them. This turns out not to be so, or at least not so in Japan, where puffer fish have been eaten since Neolithic times. The flesh of puffer fish is said to be exquisite, and the Japanese regard it as a culinary delicacy. The threat of fatal poisoning seems not to intimidate those who enjoy this delight. In fact, it is three of the most poisonous species of puffer fish whose flesh is prized most highly. The poison is concentrated in the fish's ovaries, liver, intestine, skin, and spawn. If these are painstakingly removed without contaminating the flesh, puffer fish can be eaten without fear of death. Cleaning puffer fish properly is an exacting art, and the Japanese Ministry of Health and Welfare grants special licenses to professional cooks who master it. In restaurants employing a licensed cook, puffer fish poisoning is said to be practically nonexistent. Nonetheless, one hundred to two hundred Japanese suffer puffer fish intoxication each year. About half of these victims die, typically from respiratory failure.

Although we think of tetrodotoxin as the puffer-fish toxin, the compound actually originates elsewhere. The first indication that tetrodotoxin was not peculiar to puffer fish came in the 1960s, when Harry S. Mosher, a chemist at Stanford University, found the compound in the eggs of rough-skinned newts (*Taricha torosa*). Subsequently, investigators screened a wide variety of marine organisms and found tetrodotoxin in many unrelated species, including various octopuses, crabs, algae, and starfishes, as well as in several fishes related to puffer fish. Japanese scientists carefully examined the organisms in a particular coral-reef environment, searching for the original source of tetrodotoxin. From their research, it appears that the organism responsible for producing tetrodotoxin in this reef was an *Alteromonas* bacterium. They also found several other microorganisms that produce tetrodotoxin. The toxin then moves from the microorganisms into other species through the food chain. One tetrodotoxin-resistant organism after another picks up the toxin in its diet and accumulates it for self-defense. If this food chain did not ultimately reach man, owing to the delicacy of puffer-fish flesh, we might still be unaware of tetrodotoxin and its deadly effects.

If puffer fish were not regarded as an extraordinary delicacy, doubtless no one would bother to prepare and eat them. While toxins and venoms are common among fishes, they are rare among those that we eat. Our food fishes are generally fast-swimming, carnivorous, plainly colored species that live in the open ocean. Such species do not depend on chemicals for their defense. Toxins, on the other hand, are common in slow-swimming, herbivorous fishes that inhabit reefs and other enclosed spaces. Many of these species advertise their toxicity with bright coloration or distinctive markings.

Animals presumably produce many other chemical compounds that we would find useful, and the search for them goes on all the time. Only

recently, scientists at two large American pharmaceutical companies identified several new components of the venom of a desert spider, *Agelenopsis
aperta*, found in the southwestern United States and Mexico. Unlike other
toxins that irreversibly block nerve impulses, these compounds cause immediate but temporary paralysis when tested in insects. These new spider toxins may find use as tools for biomedical research or as models for new drugs,
if investigators can learn how they work.

4

CHEMICAL WARFARE IN
SIMPLER ORGANISMS

*T*HERE IS an entire world of simpler
organisms, many of which, too
small for us to see, are better known by their effects than their appearance.
Toxic chemicals are unusually widespread in these creatures, and in the past
half-century these toxins have become indispensable in human affairs, particularly agriculture and medicine. We must now examine chemicals from
these organisms with care.

Two large groups of these simpler organisms are bacteria and fungi.
Many of these creatures are microorganisms that live by invading other species and proliferating in these hosts to form large colonies. Some of these
microorganisms attack many different species, while others invade only a
single host. Often the host organism is unaware of these microbial activities,
but certain microorganisms cause diseases that can disable or even kill the
host. Organisms that cause disease are called pathogens, and pathogenic microorganisms frequently release toxins that bring about some or all of the
characteristic symptoms of specific diseases. Several pathogens that provoke
plant diseases deploy toxins as powerful offensive weapons in overwhelming
their hosts. These compounds are chemical warfare agents that play an important part in spreading plant diseases.

Because some of these diseases strike important food crops, agricultural
scientists have long been interested in how these toxins work. They have

discovered that some toxins are poisonous only to a single plant species or even only to a particular strain of one species. Typically, these host-specific toxins are important early in the course of a disease, just as the pathogen is beginning to infect the plant. The toxin weakens the plant at this crucial time, making it easier for large numbers of the microorganism to breach the plant's defenses, enter, and proliferate.

Strangely enough, different strains of the same microorganism may produce unrelated poisons, each of which affects only one particular host. Each strain then specializes in invading a different plant. There are at least six distinct forms of the fungus *Alternaria alternata*, and each of them releases a chemically distinct toxin. Because each toxin is specific for a particular host, this one fungus is responsible for several different plant diseases. One form of *A. alternata* attacks strawberries (*Fragaria* × *ananassa*) and leads to a condition known as strawberry black spot. Another form of the fungus infests tobacco plants (*Nicotinia tabacum*) and causes tobacco brown spot, while a third form induces stem canker in tomatoes (*Lycopersicon esculentum.*) Other fungi that synthesize host-specific toxins attack oats, sorghum, and sugarcane.

Another group of pathogens releases toxins that are not host-specific. Most often, these compounds interfere chemically with vital biochemical processes common to many organisms, so they are poisonous to a wide range of plants, and sometimes animals as well. Nonspecific toxins typically facilitate plant disease in a different way from host-specific toxins. Instead of expediting the pathogen's entry into its host, most nonspecific toxins increase the virulence of the disease once the pathogen is established in the host. As the pathogen proliferates and overwhelms its host, these toxins exacerbate the resulting disease. One nonspecific toxin that works this way is a compound called tabtoxin. Plant scientists first found tabtoxin in a particular strain of the bacterium *Pseudomonas syringae* that invades tobacco and causes a disorder known as wildfire disease. Other strains of the same bacterium that also release tabtoxin strike food crops such as oats, beans, and soybeans. Because tabtoxin disables a widespread, essential enzyme, it is poisonous to all these crop plants. Nonspecific toxins from other pathogenic organisms figure in diseases of citrus fruits, almonds, and corn.

Scientists generally agree upon the roles of host-specific and nonspecific toxins in promoting plant disease. They disagree, however, on the roles

4.1 *Leaves of bean, soybean, and tomato plants, each infected with a different form of* Pseudomonas syringae. *At least two dozen additional forms of this pathogen attack other common crop plants.*

of other toxic compounds released by many microorganisms, generally because there is little experimental evidence to reveal what these toxins do. Some scientists support the commonsense view that organisms produce these toxins only because the toxins serve them as important chemical weapons. This reasoning is based on the principle that organisms devote energy only to activities that improve their chances of survival and reproduction.

Other scientists remain unconvinced by this argument, pointing out that in many cases there is really no evidence that toxins help the organisms that release them. In this view, these chemicals may have some significant function that remains unknown to us, and their toxicity may be merely fortuitous. Until compelling proof is at hand, the disagreement will probably continue, although the view that these toxic compounds are chemical weapons seems to be gaining ground.

COMBATTING HUMAN DISEASES WITH MICROBIAL TOXINS

Although we know relatively little about the ways many microbial toxins serve the microbes that create them, we can say a great deal about how these

57

compounds affect human society. For us, the most important microbial products are nonspecific toxins that destroy pathogenic microorganisms, while being relatively nontoxic to humans. Over the past fifty years, such toxins have become increasingly valuable as antibiotics and have revolutionized clinical medicine. The idea of using microbial toxins to combat human pathogens originated in 1928 with a chance observation by Alexander Fleming, a British bacteriologist at St. Mary's Hospital Medical School in London. Fleming noted that the bacteria he was growing in culture did not survive in an area on the culture plate that was contaminated with the mold *Penicillium notatum*. He pursued this observation and found that a toxin produced by the mold had killed the nearby bacteria. Fleming named his discovery "penicillin," after the mold, and went on to demonstrate that this toxin destroyed many different pathogenic microorganisms but had little effect on laboratory animals. He speculated that penicillin might have application in clinical medicine.

Fleming did not follow up his idea of applying penicillin in medicine, and ten years passed before other investigators did so. At the outset, a major impediment to testing or using penicillin was that the toxin was chemically unstable and difficult to purify without diminishing its antimicrobial activity. Accumulating a useful quantity therefore was difficult. Scientists at Oxford University, led by Howard W. Florey and Ernst Chain, solved the problems of isolating and purifying penicillin. They then tested the compound clinically and discovered that Fleming's speculation was correct: penicillin was extraordinarily effective in combatting bacterial infections in human patients.

By the time Florey and Chain had obtained these results, the Second World War was engulfing the world. Infection is a formidable problem on the battlefield, and British and American military medical officers were extremely interested in this new drug. At their urging, penicillin research was expanded, and the antibiotic was put into production as rapidly as possible. Even before large quantities were available, it was saving the lives of thousands of men wounded in battle. Penicillin was in short supply through 1943, but adequate amounts were on hand by the time of the Allied invasion of Normandy in June 1944.

The wartime development of penicillin in Great Britain and the United States was shrouded in military secrecy. After the war, the penicillin story became widely known, and the drug became available to clinicians everywhere. Penicillin was the first of many naturally derived antibiotics, and it initiated a revolution in the treatment of infectious disease.

NEW INSECTICIDES FROM MICROBES

The development of antibiotics from microbial toxins has contributed enormously to human welfare. These compounds that kill pathogens without harming humans seemed magical when they were first discovered, and their phenomenal success led scientists to consider whether microbial toxins could be turned to advantage elsewhere. One attractive possibility was the control of agricultural pests. If microbial toxins could be converted into biological insecticides, perhaps they could offer a practical solution to a problem of growing importance.

This problem concerned the serious side effects of using chemical pesticides in agriculture. Synthetic chemical compounds came into general use as insecticides in the late 1940s, and within a few years their shortcomings began to emerge. The effectiveness of chemicals started to decline as pest species evolved pesticide-resistant strains. Farmers found themselves using more and more insecticides to achieve the same results. At the same time, the environmental costs associated with using chemicals were becoming more obvious each year.

Chemical pesticides and their deficiencies became an issue of wide public concern in the United States following the publication of Rachel Carson's *Silent Spring* in 1962. As this haunting book swept the country, Americans began to appreciate the environmental price exacted by the heavy use of pesticides. The goal of higher crop yields was devastating the countryside. Plant scientists believed that insecticides based on microbial toxins would cause less damage to the environment, and might be more effective than chemicals. Biological insecticides, or biopesticides, presumably would be toxic to targeted pest species but harmless to vertebrates and beneficial insects. In addition, biologists expected pests to develop resistance to natural toxins much less readily than to synthetic pesticides. Ultimately, biopesticides were environmentally appealing because they could be used in small amounts and would be biodegradable. The pollution of land and sea, a concern with chemical pesticides, might no longer be a problem.

The prospect of biopesticides was indeed attractive and remains so today. It has inspired plant scientists for more than thirty years as they work to turn microbial toxins into practical insecticides. Some of their most promising discoveries have come from a common soil bacterium that was recognized as an insect pathogen nearly one hundred years ago. This bacterium is formally known as *Bacillus thuringiensis* and is usually called BT. There are more than two dozen varieties of BT, each synthesizing its own mixture of protein toxins in the form of small insoluble crystals. As far as we know, none of these toxins is harmful to mammals or birds.

In some cases, proteins from different varieties of BT are toxic to different groups of insects. The many varieties of the bacterium and the several different toxins create a complex situation, but agricultural scientists have turned this complexity to advantage. From one variety of BT they have developed an insecticide particularly toxic to the larvae of butterflies and moths. This insecticide can control such species as the tobacco budworm (*Heliothis virescens*), the cabbage looper (*Trichoplusia ni*), and the European corn borer (*Ostrinia nubilalis*), all major agricultural pests. Another BT variety produces proteins toxic to members of the order Diptera (flies and mosquitoes). From this latter variety has come an insecticide effective against the mosquitoes that carry malaria (*Anopheles* species), and those associated with yellow fever and dengue fever (mainly *Aedes aegypti*). A third strain of BT is particularly toxic to beetles. Some BT toxins are sufficiently specific that beneficial insect species are unaffected by them.

Pesticides containing BT toxins have been on the market since the 1960s. Most of the available products are liquid or solid preparations intended for application in standard spray equipment. Today their sales account for 90 percent of all biopesticides sold and amount to about $50 million a year. However, this is only a minute share of the market, since annual pesticide sales total about $20 billion worldwide. More than 99 percent of all pesticides sold are synthetic chemicals.

Despite their apparent scientific advantages, biopesticides based on BT toxins have found little practical acceptance. This failure in the marketplace results from diverse technical, political, and economic problems. Two straightforward but important issues are the relatively high cost and a reputation for unreliable performance. BT has complex growth requirements, making the toxins relatively expensive to produce and harvest. The toxic proteins are also biodegradable and chemically fragile, leading to the possibility of reduced insecticidal activity and inconsistent performance of commercial products.

A more general issue is the nature of the competition that BT insecticides confront. The synthetic chemicals industry is a well-established multi-billion-dollar enterprise that provides thousands of good jobs in the United States. As such, it exercises substantial political clout at every level. It is also hard at work seeking new synthetic insecticides that are biodegradable and more specific in their action.

While industrial, governmental, and academic laboratories continue their efforts to make BT pesticides a practical success, a grave new complication has appeared: pest species are acquiring resistance to BT toxins. In January 1992, plant scientists met under the auspices of the USDA's Agricultural Research Service to review evidence of this development and to propose possible remedies.

Evidence for BT resistance had been accumulating for several years. It implicated eight pest species, including such economically significant insects as the corn earworm and the Colorado potato beetle. (For comparison, more than five hundred pest species are now resistant to chemical pesticides.) Most BT resistance has shown up only in laboratory experiments, but at least one economically significant insect has already evolved high-level resistance in the field. This is the diamondback moth (*Plutella xylostella*), which is the world's most destructive insect of such cruciferous crop plants as cabbage, broccoli, and mustard. This insect has a history of acquiring resistance quickly and easily. In 1953 it was the first crop pest to become resistant to DDT. It is now resistant to every chemical pesticide used against it.

Plant scientists had believed that the nature of BT's toxic action would prevent pests from acquiring resistance to them so quickly. Pests ingest the toxin crystals as they feed. The crystals dissolve, and the toxin disrupts the membranes of cells lining the gut. This disturbance can have several different fatal consequences. Scientists had presumed that the existence of multiple lethal pathways would preclude simple genetic changes in pest species from blocking the toxin. It now appears that this complex mode of action presents many points at which differences associated with a simple genetic change may protect the pests instead.

At the 1992 conference, plant scientists endorsed several strategies to combat BT resistance. One simple suggestion was for farmers to practice alternation or rotation of BT toxins with other pest control methods. Another idea was for manufacturers to expand the number of BT toxins available for use against a specific pest. Current commercial preparations typically contain a single toxin. One reason for rapidly acquired resistance to commercial BT toxins may be that pests are repeatedly exposed to single-toxin preparations. In contrast, these same pest species do not become resistant to the complex mixtures of toxins produced by the several varieties of BT they encounter in nature.

Several other strategies for combatting resistance to BT came out of the conference, but none of them offers an easy answer to this unwelcome development. If acquired resistance is to be overcome, plant scientists, manufacturers of BT pesticides, and farmers will have to implement one or more of these strategies on a large scale.

BT biopesticides have scarcely begun to fulfill their promise. After thirty years on the market, serious objections still cloud their future. Nevertheless, in late 1992 agricultural scientists reporting in *Science* on the conference held earlier that year characterized BT toxins as the "most scientifically, environmentally, and sociologically acceptable pest suppression tools of this century and possibly the next." Trying to make them work is certainly a worthwhile goal.

FIGHTING NEMATODES

Although Fleming had discovered penicillin by accident, the search for new antibiotics after the Second World War was a carefully organized endeavor. The success with penicillin implied that other microbial toxins might be clinically useful, and pharmaceutical companies in Europe and the United States launched screening programs to identify candidate antibiotics. Within a few years, investigators recognized that a particularly rich source of promising compounds was a group of fungus-like soil bacteria called actinomycetes. These organisms quickly became prime targets for future screening, and they ultimately yielded hundreds of new drugs. From just one genus of actinomycetes (*Streptomyces*) have come such familiar antibiotics as actinomycin, neomycin, streptomycin, and oxytetracycline.

Actinomycetes produce many toxins, but only a few of these are suitable for development into drugs. The most common shortcoming is excessive toxicity to mammals. For a toxin to be a candidate antibiotic, there must be a considerable difference between the dose needed to kill a pathogen and the dose that can harm the patient. Many toxins are simply too poisonous to mammals for use in human or veterinary medicine. In addition, some toxins are chemically too unstable, and others are produced in amounts too small to be usable. By the 1970s, scientists in the laboratories of pharmaceutical companies were designing screening procedures to identify toxins for a variety of specific clinical applications.

One new screening test, or assay, used in the United States at the Merck Sharp & Dohme Research Laboratories in 1975, was designed to search for drugs to control parasitic nematodes, particularly those that infect farm animals. Nematodes, or roundworms, are familiar to gardeners as soil organisms that ravage plants. They are a large phylum of slender, shiny worms with unsegmented bodies that range from less than 1 millimeter to over 8 meters in length. These worms have parasitized almost all species of plants and animals, and they cause a variety of common diseases in dogs, pigs, cattle, and other domestic animals. About fifty species of nematodes live in humans, most of them passing quiet lives unnoticed by their hosts. A few of them create real medical problems, however, and provoke such human diseases as elephantiasis, hookworm, and trichinosis.

Nematodes and other parasitic worms are organisms quite unlike microbial pathogens. Not only are they larger, multicellular creatures, but their bodies run on biochemical machinery that is rather different from that used by microbes. For this reason, most antihelmenthic agents, or drugs effective against nematodes, are chemically different from the drugs used to fight bacterial and fungal diseases.

Merck's new procedure for finding antihelmenthic agents allowed scientists to test microbial toxins directly against nematodes living as parasites in laboratory mice. The investigators obtained the compounds to test by growing different microbes in a nutrient broth. They then mixed the liquid containing the products of microbial growth, including any resulting toxins, with mouse feed. They dried this mixture of feed and toxins and fed it to mice infected with a nematode called *Nematospiroides dubius*. This worm is a common parasite of wild rodents, and it serves as a convenient organism for laboratory investigations. It lives in its host's intestine and produces eggs that are excreted with the host's feces. Worm-infected mice ate the special diet containing toxins, and the scientists noted the weight and general health of the mice day by day. They also examined the mice for nematode eggs and worms. From these observations they could determine how the microbial products in the mice's feed affected both the mice and the nematodes. They looked for toxins that eliminated the nematode infection without harming the mice.

The Merck investigators applied this assay to cultures of microbes from soil samples obtained from various parts of the world. Microbial toxins that killed the nematodes were nearly always toxic to the mice as well. Such compounds were too toxic to hold any promise as drugs and were examined no further. One culture, however, seemed to merit more careful exploration. This culture bore code number OS-3153 and was one of many that Merck had obtained from the Kitasato Institute in Tokyo. Culture OS-3153 contained microorganisms from soil dug up in Kawana, on the Japanese coast about 60 miles southwest of Tokyo. Mice whose feed contained products from OS-3153 maintained nearly normal weight and were free of nematode eggs and worms. Something from OS-3153 killed the worms but did not harm the mice. Further experiments showed that, whatever this something was, it was toxic to nematodes at a very low concentration. Culture OS-3153 was the single positive lead for an antihelmenthic agent that the Merck investigators discovered in examining over one hundred thousand cultures.

The Merck scientists now focused their attention on a detailed examination of culture OS-3153. Microbiologists found that the organism in OS-3153 was a previously undescribed member of the genus *Streptomyces*. They named it *Streptomyces avermitilis*, which they translated as "the *Streptomyces* capable of separating from worms." Other experts modified the conditions for growing *S. avermitilis* in the laboratory to maximize production of the antihelmenthic drug. Chemists purified the drug and determined that it was a mixture of eight closely related chemical compounds. They named these compounds avermectins and worked out their chemical structures.

The chemists also prepared a new active compound in the laboratory by slightly modifying the chemical structure of one of the avermectins. Modifying natural toxins to create compounds with improved properties is a common maneuver in drug development. Often minor changes in the chemical structure of a compound can impart increased stability, reduced toxicity to humans, greater or different specificity of effect, or other desirable properties to a candidate drug. The resulting compounds, which are neither totally natural nor totally synthetic, are usually called semi-synthetic compounds.

The chemists gave the name "ivermectin" to their modified version of avermectin. Tests showed that ivermectin was more stable and safer to use than any of the compounds obtained directly from *S. avermitilis*. In time, ivermectin became the preferred form of the drug for practical application. For convenience, we can use "ivermectin" to refer to the active drug from OS-3153.

Biologists tested ivermectin for activity against a wide variety of organisms. They wanted to find out how effective it was against nematodes other than *N. dubius*, and also whether it was toxic to pathogenic bacteria, fungi, insects, and other target organisms. Their candidate antibiotic proved to be among the most potent therapeutic agents known. Ivermectin was extremely toxic to eight different families of nematodes, as well as a powerful poison for insects and mites. The Merck investigators next examined the mode of action and toxicology of ivermectin, and finally undertook clinical trials of the drug for specific applications. These extensive investigations required years of effort by many hundreds of people. Ivermectin emerged from Merck's long scrutiny as a versatile and effective antibiotic.

By 1990, ivermectin was in broad clinical use. The screening procedure involving nematode-infected mice had yielded just what it was designed to yield. Ivermectin is effective against nematode parasites in cattle, pigs, sheep, and horses. It has also become an important agricultural pesticide for use against mites and insects, both to control those that attack farm animals and also against herbivorous species that feed on economically important crop plants. It shows promise for managing one of the common kinds of cockroach (*Blatella germanica*), as well as red imported fire ants (*Solenopsis invicta*), which are significant pests across the southeastern United States.

Despite its toxicity, ivermectin poses only a limited threat to the beneficial insects and mites that have their own critical roles in agriculture. Many beneficial species feed on arthropod pests and keep their numbers under control. Others pollinate agricultural and horticultural plants. Most broad-spectrum commercial pesticides disable vital biochemical systems common to many kinds of organisms, and as a consequence they annihilate helpful and harmful species indiscriminately. Ivermectin, in contrast, is more selective than broad-spectrum pesticides and spares many beneficial organisms.

Owing to a favorable combination of properties, ivermectin discriminates between helpful and harmful species. When ivermectin is sprayed on a plant, it passes quickly into the leaves. The drug retains its toxicity inside the leaves and poisons herbivorous pests when they feed on the plant. Because beneficial arthropods do not feed on the plant, ivermectin within the leaves does not endanger them. If a beneficial arthropod lands on a sprayed plant or crawls across a sprayed leaf, it comes into contact only with the small amount of ivermectin remaining on the plant's outer surface. Moreover, this surface residue remains toxic only briefly, because ivermectin is light-sensitive. Sunlight soon degrades the residue exposed on the plant's surface and renders it nontoxic. In consequence, species that do not feed on ivermectin-treated plants have little risk of being poisoned.

These favorable characteristics have made ivermectin a successful and profitable product in agriculture and veterinary medicine. It has performed exceptionally well in a variety of markets under many trade names and in many formulations. By 1987, ivermectin was Merck's second largest-selling drug, and its sales in 1992 were estimated at half a billion dollars. Yet ivermectin's remarkable commercial success may be eclipsed by a role of even greater significance: in addition to dominating plant and animal diseases, ivermectin holds a unique position in human medicine and could affect the lives of millions of people.

Potentially, ivermectin can control river blindness, or onchocerciasis, a parasitic disease endemic to tropical portions of Africa, the Near East, and the Americas. The affected areas lie in thirty-five developing countries and are inhabited by 85 million people, about 18 million of whom are afflicted with the disease. One to 2 million of those afflicted are visually impaired or blind, and many others suffer from a disfiguring skin disease. More than half of the people living in heavily infested areas will be blind before death. The life expectancy of those blinded by onchocerciasis is about one-third that of their sighted peers.

To appreciate how ivermectin can alleviate this wretched affliction, we should sketch briefly the natural history of onchocerciasis. The disease begins when a human is bitten by a minute blackfly infected with a parasitic nematode called *Onchocerca volvulus*. The blackflies breed in running water, so people living near rivers are particularly at risk. In West Africa, the black-fly that carries onchocerciasis is *Simulium damnosum*. Other *Simulium* species are the carriers elsewhere. When an infected fly bites a human, it deposits larvae of the nematode in its victim. Within about a year in the human body, these larvae grow into adult worms.

The adult worms collect as a tangled mass in nodules in various parts of the human host's body. They remain in their nodules and can live for as long as eighteen years. The nodules can be painless, but the worms reproduce and give rise to immense numbers of immature forms called microfilariae. The microfilariae migrate from the nodules and wander throughout the host's body. They are only a few tenths of a millimeter in length, but a single human host may harbor more than 200 million of them. When a blackfly bites an infected human, it picks up some of these microfilariae. During about a week in the fly's body the microfilariae grow into larvae and thus complete the nematode's life cycle. The infected fly carrying larvae is now capable of transmitting them to a new human host and starting or aggravating an infection of onchocerciasis.

The debilitating effects of onchocerciasis arise from the huge number of microfilariae borne by a human host. Blackflies that bite an infected human pick up only a negligible fraction of the millions of microfilariae generated by the resident adult nematodes. Most of the microfilariae end their wanderings by concentrating in the host's skin or eyes, where they eventu-

4.2 Three species of blackflies that cause river blindness in Guatemala.

ally die. In human skin, dead microfilariae cause a tormenting itch. (The specific name of the blackfly *Simulium damnosum* comes from the damnable itch that accompanies onchocerciasis.) The microfilariae that accumulate and die in the eye have a much worse consequence. They impair vision, and in time they cause total blindness. In addition, skin lesions and secondary infections accompany onchocerciasis, and weakened victims of the disease are susceptible to other ailments.

In the past, there was no satisfactory direct treatment for this disease. Surgical removal of the nodules is fairly successful but offers no practical solution, because nearly all the victims of onchocerciasis are impoverished rural people, many of whom live in geographically isolated areas. Ivermectin is the first drug to offer real hope of combatting the disease. With as little as a single dose per year and only mild, transient side effects for the patient, ivermectin stops adult worms from generating microfilariae, and it kills those microfilariae already present in the host.

The drug does not kill the adult nematodes and so cannot cure the condition. It cannot reverse the blindness or other debilitating effects of onchocerciasis once they are established, but it does suppress the buildup of microfilariae that produces these effects. Ivermectin also slows the transmission of onchocerciasis by wiping out the human reservoirs of microfilariae that would otherwise infect more blackflies. Eradicating microfilariae in humans disrupts the parasite's life cycle because, as far as we know, the adult nematodes live and generate microfilariae only in humans.

Ivermectin's potential for slowing the spread of onchocerciasis is evident from a large clinical trial carried out a few years ago. The test population was fourteen thousand people living on a rubber plantation in a forested area of West Africa. These people were heavily infected with onchocerciasis. Twenty-six percent of those five years old and 86 percent of those twenty years old tested positive for the disease before the trial began. Each individual in the test population received an oral dose of the drug three times a year for two years. At the end of this period, the incidence of onchocerciasis in children seven to twelve years old had dropped 45 percent. This dramatic improvement depended on treating the entire population throughout the two-year experiment, so that newly emerged blackflies could no longer become infected with microfilariae.

The remarkable efficacy of ivermectin in treating onchocersiasis emerged during Merck's extensive testing of the drug. By 1980, Merck scientists had established ivermectin's activity against nematodes related to *Onchocerca volvulus*, and they arranged to test the drug against the human parasite. A small field trial in Africa the following year was encouraging. Trials over the next few years revealed that, for onchocerciasis, ivermectin was truly a wonder drug. In 1987 Merck & Company announced that it would offer ivermectin free of charge for treatment of onchocerciasis for as long as it was required. A corporate spokesman noted that the drug "was needed only by people who couldn't afford it." This unique offer surprised the pharmaceutical industry and committed Merck to a considerable future expense.

The ivermectin program is unprecedented in its projected scope. The goal of controlling onchocerciasis throughout the areas of risk requires that millions of people receive a dose of ivermectin once a year for about fifteen years, a total of well over 1 billion individually administered doses of the drug. If this can be done, there is a good chance not only of controlling onchoceriasis, but of actually eradicating the disease. It is not yet clear whether this goal is realistic. The program for distributing ivermectin got under way in 1988, and by the end of 1989, about four hundred thousand doses of the drug had been dispensed. By 1991, the number had reached 3.2 million. Ivermectin has now reached between 5 and 10 percent of the population at risk for onchocerciasis. The logistical and political problems in delivering the drug are enormous, and they can only multiply as the program progresses. Each year, public health officials move to treat less and less accessible populations in areas where facilities are primitive at best and travel is barely feasible. Ultimately, however, success in conquering onchocerciasis would tangibly improve the lives of millions of people who have very little in this world. It would be an admirable achievement that no one could have imagined when the soil of culture OS-3153 was first dug from the earth.

THE PUZZLING DEATH OF THE ENGLISH TURKEYS

Antibiotics are obviously some of the most beneficial compounds from nature, but there are also microbial toxins that are significant to us because they are harmful. There is an important group of toxins, largely from fungi, that are dangerously poisonous to human beings and to other animal species. These lethal toxins appear to be chemical warfare agents directed against us, although human beings are not their natural target. Because these compounds endanger farm animals as well as human beings, both agricultural scientists and public health officials devote serious attention to understanding and controlling them. Some of these toxins have been familiar for centuries, but the most thoroughly investigated member of the group is aflatoxin, a fungal toxin that was discovered only in 1960.

Aflatoxin is exceedingly toxic and carcinogenic, and it turns up in human food and animal feed all over the world. It is an exceptional toxin, and it has an exceptional history. Unlike most chemicals, aflatoxin came to light as much through laborious detective work as through scientific research. Both efforts began with the puzzling deaths of tens of thousands of English turkeys (*Meleagris gallopavo*).

During the summer of 1960, a strange disease suddenly struck English turkey farms. Young birds became lethargic, lost their appetites, and died within a week. Postmortem examination revealed that they died with inflamed intestines and swollen kidneys. The liver was frequently involved, and occasionally the pancreas. Birds that appeared healthy in the evening were sometimes found dead the next morning. One expert, visiting a farm where eight hundred young turkeys had already succumbed, discovered the remaining birds dull and lifeless. As he watched for half an hour, more than a dozen of them died. Oddly, some farmers reported the disease in one group of birds but not in others kept in an adjacent pen.

Veterinarians could find no microorganism or virus responsible for the disease, nor were they able to transmit the disease from sick birds to healthy ones. They ruled out the known turkey diseases, including some previously encountered in the United States but not in Britain. With no causal agent at hand, the condition became known as turkey "X" disease. It soon spread to ducks and pheasants. Government investigators considered the possibility that the birds were being poisoned. They sought the advice of experts on poisons from several outside laboratories and agencies, including Scotland Yard. Technicians carried out extensive laboratory tests on the birds' feed, checking for the presence of a large number of natural and synthetic poisons. All their tests were negative. Meanwhile, birds continued to die, as many as five thousand in a single flock. By late August, about five hundred farms had been affected. Before the epidemic was over, turkey "X" disease killed more than one hundred thousand young turkeys.

Agricultural investigators noted that turkey "X" disease did not strike in

Northern Ireland, Wales, or Scotland, and that very few cases appeared in the north of England. The problem was concentrated in the south of England, and nearly 80 percent of the affected farms were within one hundred miles of London. Following up this clue, investigators searched for some critical local factor in the birds' deaths. Success came when a farm-by-farm check established that the feed given to all the dead birds had come from one of two London mills. The feed from the two implicated mills contained peanut meal, a common constituent of poultry rations that was generally imported into Britain from Nigeria or India. The investigators learned, however, that these two London mills recently had been using peanut meal from a new and different source. This meal had come from Brazil and had arrived in England aboard the SS *Rosetti* in late 1959.

Government scientists obtained a large sample of this shipment, which became known as "Rosetti meal," for laboratory examination and feeding tests. They quickly found that young turkeys that were fed Rosetti meal developed the symptoms of turkey "X" disease and died. Rosetti meal apparently contained a powerful poison that had gone undetected in the earlier tests. To solve the mystery of turkey "X" disease, the investigators now had to find and identify this novel poison. Such a task could be a chemist's nightmare. If the poison was very toxic, there need not be much of it present. If very little poison was present, it would be more difficult to pinpoint. In addition, only two things were known about this poison: it was not detected in any standard laboratory test, and it killed birds. The chemists would need more information before they could hope to isolate and identify the toxin.

The most important additional information came from examining Rosetti meal under a microscope. The meal proved to be heavily contaminated with a fungus. It was easy to see bits of dead fungus in the peanut meal, but because the fungus was dead, identifying it was difficult. Mycologists (scientists who study fungi) finally concluded that the contaminant was a common mold known as *Aspergillus flavus*. When they grew a particular strain of this mold on sterile, nontoxic peanuts, it produced a previously unknown toxin. This new toxin elicited the symptoms of turkey "X" disease in young birds. Now that they knew what to look for, chemists were able to find the same toxin in Rosetti meal. It received the name "aflatoxin" in recognition of its origin in *A. flavus*.

Later studies showed that only certain strains of *A. flavus* and a closely related fungus, *A. parisiticus*, produce aflatoxin. The chemical composition of the toxin depends on the strain of fungus producing it and on how the fungus itself is grown. Chemists eventually obtained a dozen different compounds from various samples of aflatoxin. These compounds were chemically quite similar to one another, but they differed greatly in their toxicity to birds and mammals. About 1 milligram of the most toxic compound was sufficient to kill a duckling or young turkey. The long-term toxicity of low levels of aflatoxin was also quite high. As little as 1 percent of the contaminated Rosetti meal in the birds' diet led to severe liver damage.

From these experiments, investigators realized that aflatoxin was extremely toxic and that even in minute doses its cumulative effects could be fatal. Agricultural and food scientists collected samples of peanuts (*Arachis hypogaea*) and peanut meals from hundreds of sources and tested them for aflatoxin to determine how widespread this toxin was. They found that contamination was variable and sporadic, but samples from thirteen different countries were poisonous. Aflatoxin ultimately turned up in peanuts from every country that grows them. Additional tests demonstrated that certain strains of *A. flavus* and *A. parisiticus* also release aflatoxin when they are grown on foods and grains other than peanuts.

With these findings in hand, scientists appreciated that aflatoxin could have easily turned up earlier in food or animal feed in many parts of the world. It seemed unlikely that turkey "X" disease was the first instance of large-scale aflatoxin poisoning. On scanning technical descriptions of unexplained animal diseases reported over the preceding thirty years, scientists identified several outbreaks in Europe and the United States with the characteristics of aflatoxin poisoning. Aflatoxin was apparently nothing new, but earlier incidents had been left unexplained.

There have also been cases of aflatoxin poisoning affecting humans since the appearance of turkey "X" disease in 1960. Probably the worst of these was an outbreak in India in 1974, when more than one hundred people died from eating moldy grain. There have also been less dramatic, long-term effects. A recent study found that two-thirds of the people in the Chinese city of Qidong showed signs of aflatoxin poisoning. About sixty people per hundred thousand in Qidong die of liver cancer each year. This is the highest rate of any area in China and is many times the rate in Western countries.

As a consequence of turkey "X" disease, many veterinary scientists, chemists, and mycologists had inquired into aflatoxin and its biological effects. Interest quickly expanded to a much wider scientific community because the British Medical Research Council Laboratories soon reported that several components of the toxin were extremely carcinogenic to laboratory animals. One component, called aflatoxin B_1, was unlike any carcinogen seen before. Aflatoxin B_1 induced liver cancer in rats when it was present in their diet at a level as low as fifteen parts per billion. This is an amount of aflatoxin B_1 weighing as much as one grain of table salt distributed evenly in about five pounds of rat food! In fact, aflatoxin B_1 is probably the most potent natural carcinogen that is known.

Aflatoxin assumed additional importance in the mid-1970s when the World Health Organization (WHO) decided that peanut meal was a good source of protein for undernourished children. Public health officials then had to ensure that they were not supplying large numbers of children with a highly carcinogenic toxin in their diet. All in all, aflatoxin has received more attention than any other fungal toxin.

TOXIC TIDES AROUND THE WORLD

Although no one suspected the existence of aflatoxin until 1960, a different toxic menace has been familiar around the world for centuries. This menace comes from the sea, and it threatens food supplies just as aflatoxin does. Over the years, many different peoples must have encountered and dealt with seaborne poisons. The long experience of the coastal Indians in western North America is probably typical.

Northwestern coastal Indians make their homes along the Pacific from Puget Sound to Alaska. Regional peoples, such as the Nootka and the Coastal Salish, traditionally lived on salmon and other fish, and supplemented this diet with shellfish. All along the northern Pacific coast, there are immense banks of discarded shells, some dating back more than five thousand years, that attest to the antiquity and success of this way of life. Mussels and clams were plentiful, and a child could collect them with only a stick for a tool.

The Indians did not gather and eat shellfish all year round, however. They knew that when the tide became discolored in July or August, it brought poison to their shores, and that eating mussels and clams at this time was dangerous. Well into the twentieth century, these Indians watched each summer for the local waters to turn murky and whitish. The fisherman who saw the first murky tide of the summer alerted other villagers, and they stopped gathering shellfish at once. For the next two or three months, the Indians avoided eating mussels and other marine invertebrates.

The water usually remained discolored with toxin for a month or more, and after it cleared, the Indians waited another month before starting to eat shellfish again. During periods of polluted water, they regarded all marine invertebrates as dangerous but believed that the California mussel (*Mytilus californianus*) and the common blue mussel (*Mytilus edulis*), were particularly toxic. Because salmon were unaffected by the poison, the Indians suffered little hardship from the pollution, as long as they avoided invertebrates.

Early European visitors to the Pacific Northwest were unaware of the danger that the native peoples had long since learned to avoid. On Captain George Vancouver's voyage of discovery in 1793, one sailor in an exploration party died after eating mussels gathered on the beach for breakfast. A few years later, the Russian Baranov expedition lost one hundred men to toxic mussels gathered near Sitka in southeastern Alaska.

Discolored waters like those in the Pacific Northwest appear all over the earth and are known as red tides. They usually occur in salt water and are frequently somewhat toxic. They result from a sudden proliferation, or blooming, of one or more species of colored aquatic microorganisms that are collectively called plankton. The actual color of a red tide depends on the organism. In a bloom, the concentration of plankton can reach over 1 billion organisms per liter of water (about one hundred thousand organisms per drop). This amount is about one hundred times greater than the concentration necessary to make a bloom visible, and also is much higher than necessary to block light and to deplete oxygen from the water. Even in a nontoxic bloom of plankton, many creatures perish from lack of light or oxygen. As fish and other organisms die and decompose, they add to the pollution. Whole communities of marine creatures can be destroyed in this way.

Toxic red tides are usually blooms of dinoflagellates, which are single-celled planktonic creatures with characteristics traditionally associated with both plants and animals. (Botanists study them as algae, while zoologists claim them as protozoa. Modern classification schemes avoid this uncertainty. Algae, protozoa, dinoflagellates, and some other microorganisms form a new kingdom called Protoctista, separate from both plants and animals.) For us, the dinoflagellates' most significant characteristic is that some of them release very toxic compounds. They may synthesize these compounds themselves, or perhaps they carry unidentified bacteria or viruses that do so. As with most other microbial toxins, scientists disagree over how useful the compounds are to the organisms that release them.

Whatever other roles these toxins may have, they certainly bring harm to human communities. The polluted waters and poisoned marine life that accompany red tides are economically devastating for fishermen and others whose lives depend on the sea. Salt spray and mists from toxic blooms can travel 30 or 40 miles on the wind and provoke respiratory distress in thousands of unsuspecting people. Worst of all, the toxins get into the marine food chain and endanger public health. This of course is what threatened the Northwestern coastal Indians and is why they avoided eating shellfish when the water was discolored.

The problem of toxins in the food chain arises because, as we have seen before, organisms often ingest poisons to which they are resistant and then accumulate these compounds as chemical defenses against their own predators. Fish generally do not sequester these toxins. On the other hand, even a relatively weak bloom of dinoflagellates can render mussels dangerously toxic to humans. As the Indians recognized, mussels sequester red-tide toxins very efficiently. Tainted shellfish can cause gastrointestinal pain, respiratory paralysis, and cardiovascular collapse in a reaction known as paralytic shellfish poisoning (PSP). Death from PSP is rare in humans, but as few as three fried clams can be lethal.

Chemists have worked out the structures of several chemically diverse red-tide toxins, and they are surprisingly complicated. The best-studied compounds from dinoflagellates include two families of toxins known as the saxitoxins and the brevetoxins. Both groups are potent neurotoxins, although they work in different ways. Saxitoxins are chemically somewhat similar to tetrodotoxin, the puffer fish poison. Both saxitoxin and tetrodotoxin bind to nerve membranes and make them impermeable to sodium ions. (An ion is an atom or group of atoms that carries an electrical charge.) This binding effectively disrupts nerve impulses, which depend on the controlled movement of ions across the membrane.

The brevetoxins are chemically quite different from the saxitoxins; in fact, their chemical structures are altogether unlike any previously known compounds and are extraordinarily complex. Brevetoxins function by opening channels in the nerve membrane to the free passage of sodium ions. This is just opposite to the effect of saxitoxin, but this free passage also interferes with the controlled flow of ions across the membrane, and so prevents the transmission of nerve impulses. Neuroscientists do not yet understand how any of the red-tide toxins works on a molecular level.

Although red-tide toxins are chemically interesting and useful to neurophysiologists as research tools, there is nothing else favorable to say about red tides. It is particularly distressing that they seem to become more frequent and more severe with the passage of time. Toxic blooms of the dinoflagellate *Ptychodiscus brevis*, the source of the brevetoxins, have been common off the Gulf coast of Florida for years, but in 1987 and 1988 for the first time they spread around the tip of Florida and extended up the Atlantic coast to North Carolina. The local shellfish industry sustained $25 million in damage, and at least forty-eight people suffered from PSP. In the autumn of 1991, a toxin called domoic acid appeared for the first time in anchovies off the California coast and in razor clams in Oregon and Washington. West Coast shellfish and crab fisheries had to be closed. This toxic bloom was especially frightening because a few years earlier on the other side of the continent, this same compound had killed three people on Prince Edward Island and had left several others with apparently permanent loss of short-term memory. In 1992, saxitoxin appeared for the first time in Alaskan dungeness crabs.

Reports at international meetings in 1987 and 1989 made it clear that red tides are a global problem. Scientists described incidents off the Atlantic coast of Spain and Portugal, in Rhode Island's Narragansett Bay, off Japan and southwestern India, in Guatemalan and Chilean waters, and in many other places. Increasingly, investigators find red tides in waters that have historically been free of them, and toxins appear in organisms previously considered to be harmless. The public health danger has consequently worsened. Worldwide, there have been about one thousand deaths from shellfish poisoning since the mid-1980s.

Many marine scientists believe that the global rise in red tides goes hand in hand with a rise in marine pollution. Both untreated sewage and agricultural runoff containing pesticides and fertilizers are rich in nitrogen and phosphorus. As more pollutants reach the oceans, the concentrations of nitrogen and phosphorus in the water go up. Long-term data from several sites show that an increase in nitrogen and phosphorus in local seawater coincides with an increase in red tides. Marine scientists also have evidence that changes in the ratio of nitrogen to phosphorus in the water stimulates some species of plankton to produce more toxins.

If red tides flourish in polluted waters, it is not surprising they are on the rise in the United States. As our population has grown over the past thirty years, pollution in American waters has increased substantially. The United States coastal population was about 80 million in 1960 and is expected to reach 127 million by 2010. Population growth means more pesticides, more sewage waste, and more pollution. Not only do red tides proliferate, but shellfish beds and fishing grounds close. The oyster harvest in Chesapeake Bay is now less than 1 percent of its level one hundred years ago.

No solution to the menace of red tides is at hand. One new possibility for fighting them comes from marine scientists at the University of Texas. They have found that certain marine viruses infect planktonic organisms and reduce their growth by as much as 80 percent. Perhaps these viruses can be used to fight red tides, but it is too soon to know whether this proposal is realistic or what its side effects might be. A group of scientists representing forty-four countries met at Newport, Rhode Island, in late 1991 and endorsed a new plan to investigate toxic blooms and their increasing frequency. This undertaking may lead to new recommended solutions, but a long-term reduction in red tides and PSP may well require more than scientists alone can provide. In addition to more information and new methods from science, the solution may demand profound political and economic decisions concerning how we use the earth's oceans.

CHAPTER

5

MEDICINES AND DRUGS
FROM PLANTS

*L*ONG BEFORE scientists began investigating nature's chemicals and
the living creatures that synthesize them, human beings were making extracts, tinctures, and other preparations from plants and animals. Thousands of years before the discovery of antibiotics, some of these natural
preparations served as medicines to relieve pain or heal sickness. Others
were poisons useful in hunting and warfare. A few acted as mind-altering, or
psychotropic, drugs and found use in religion and recreation.

Altogether, these preparations had an enormous range of properties.
One was a deadly poison from frog skin, and another was a fever remedy
from tree bark. Poppies furnished a potent pain reliever, and a certain
mushroom furnished a powerful hallucinogen. Concoctions of the most diverse sorts were common for millennia, ages before there were any biologists or chemists, or any concept of molecules and chemical compounds. All
over the earth, people actively searched the natural world for beneficial
preparations. Their success generally depended on trial and error, luck, and
serendipity.

Modern scientists have frequently analyzed preparations from earlier
times and identified the chemical compounds that are their active principles. In the light of present knowledge, most of the active compounds are
probably plant or animal chemical warfare agents. This is a reasonable

guess, but it is only a guess because knowledge in this area is surprisingly limited. As in the case of microbial toxins, we simply do not know the role of most of these compounds in nature. Human beings have been handling some of these compounds for centuries, but typically no one has investigated why they are present in the plants and animals that synthesize them.

The largest group of preparations derived from living organisms are medicines, and historically, most of these have come from plants. Civilizations throughout the world have been turning plants into medications for thousands of years, and they continue to do so today. In the centuries before modern medicine, synthetic chemistry, and the pharmaceutical industry, virtually all medicines came from plants. Today, more than 3 billion people outside the developed world depend on such remedies. Traditional medicine that relies heavily on plant and animal products is still practiced throughout China, often alongside modern Western techniques. In addition, Western medicine exploits medicinal species with considerable success. Laboratory investigators follow up leads from the traditional medications of many cultures, and they also screen previously unexamined plants and animals in the search for new active compounds. Research of the latter sort is responsible for Taxol, a compound from the stem bark of the Pacific yew (*Taxus brevifolia*) that has been much in the news as a critical anticancer drug.

The tradition of obtaining medicines from plants is sufficiently ancient that its origins are lost in prehistory. In fact, our species is not necessarily the only one to use plants as medicines. The earliest information suggesting human awareness of medicinal plants comes from pollen microfossils and other archeological evidence found in a cave in far northern Iraq. This cave was once the site of a primitive funeral. About fifty thousand years ago, a group of Neanderthals (considered by many scientists to be a subspieces of our species, *Homo sapiens*) gathered here to bury one of their dead. They prepared a grave and decorated it with local flowers carefully arranged in bunches. Their choice of flower species for this occasion is intriguing. Seven of the eight species they chose are used as medicinal plants by the current inhabitants of this part of Iraq. Some of the flowers, such as yarrow (*Achillea* species) and hollyhock (*Althea* species), are widely dispersed and are in use today as medicinals in other parts of the world. We have no way of knowing whether the Neanderthals recognized the medicinal properties of their burial decorations, but it is striking that, out of the many species available to them, they chose these particular flowers to honor the deceased. Anthropologists believe that early peoples were well acquainted with the attributes of the plants and animals among which they lived, and that hunting-and-gathering communities depended on this knowledge to survive.

OPIUM, THE OLDEST MEDICINE

For many thousands of years after this Neanderthal burial, there is no clear archeological or historical indication that humans were using plants as a source of medicines, poisons, or other active preparations. Then, sometime before the beginnings of written history, human beings discovered how to extract opium from the opium poppy (*Papaver somniferum*). Opium is both the oldest pain reliever and the first preparation used to alter consciousness. At various times throughout history it has been the world's most widely used medicine. It is still regarded as unparalleled in its ability to allay pain and impart a sense of well-being.

Human use of opium may have begun as long ago as 4000 B.C. At that time (and perhaps even a thousand years earlier), there were tribes living beside shallow lakes in what is now northern Switzerland. Archeologists have found such large numbers of fossilized opium poppy seeds around these lake dwellers' settlements that they have concluded that poppies were a crop plant for these Neolithic people. They do not know, however, why the lake dwellers raised poppies. The earliest written records are much more recent, and by that time poppies were the source both of opium and of poppy seed for food. The lake dwellers may have cultivated poppies for their seeds or for opium or for both. Unfortunately, we have no way of knowing precisely what use they made of their poppy crop. Also, there is no reason to suppose that they were the only Neolithic people to have opium poppies. It is more likely that the boggy conditions around the lake dwellers' settlements efficiently preserved and fossilized poppy seeds there, but that elsewhere organic matter routinely decomposed and disappeared. Poppies may have grown over large areas of temperate Europe in Neolithic times.

The first compelling evidence that humans had learned to make opium appears twenty-five hundred years later, in approximately 1500 B.C. Opium comes from the seed pod, or capsule, that develops after the poppy flower drops its petals. If the surface of this capsule is slit before it dries out, a milky liquid oozes out. On exposure to the air, this liquid darkens and hardens to a yellow-brown solid that can then be scraped off the capsule. This solid is crude opium. By 1500 B.C., peoples of the eastern Mediterranean were familiar with opium and probably used it for both religious and medical purposes. From Crete, there is a Minoan figurine of this period that represents a goddess with very lifelike poppy capsules sprouting like horns from her forehead. In this figurine, the capsules have been slit with just the

5.1 A libation vase decorated with a Minoan "poppy goddess." Minoan religious ceremonies also made use of the seeds of the opium poppy.

sort of incisions commonly made for collecting opium. Similar archeological evidence suggests that the Egyptians were also familiar with opium at about the same time.

By about 20 B.C., opium was well established as both a painkiller and a religious accessory in classical Greece and Rome. Throughout the imperial period, the Romans used opium-laced incense in sacrificial ceremonies. Recipes for preparing opium from poppy capsules survive from the first century, and Galen, the second-century Roman physician, was an enthusiastic proponent of opium in medicine. By the end of the second century, the emperor Severus had decreed that opium should be made available for common use. Subsequent European cultures followed this Roman practice and opium remained rather freely available as a useful medicine for centuries.

In the seventeenth century, an alcoholic tincture of opium, administered in whisky or rum, became the accepted medication in readying patients for the rude surgery of the day. Two hundred years later, opium reached its golden age. At the beginning of the nineteenth century, the German pharmacist Friedrich Sertürner succeeded in purifying the major active constituent of opium, which is the chemical compound we now know as morphine. Opium was a fashionable social drug, much popularized by the publication in 1821 of Thomas De Quincey's *Confessions of an English Opium Eater.* (English "opium eaters" did not actually eat opium but dissolved it in sherry, added some herbs for flavor, and quaffed it down.) Britain was consuming 60,000 pounds of opium a year, and in the United States, ordinary citizens bought their opium at the local general store. By mid-century, morphine had replaced opium as the mainstay of European and American medical practice.

American patent medicines could contain almost anything at that time, and opium was an attractive ingredient because it alleviated so many common complaints. The contents of these medicines remained secret, and some opium-containing nostrums were promoted as cures for opiate addiction under such names as Opacura and Denarco. An Opacura advertisement in *The Cosmopolitan* magazine promised, "Opium, morphine, laudanum and liquor habits painlessly cured by Opacura . . . As soon as the opium or liquor is expelled from the system the desire is gone." Many citizens were gratified to find that they could indeed give up opium if they took their Opacura regularly.

All this changed early in the twentieth century. After 1906, the requirements of the Pure Food and Drug Act in the United States effectively destroyed the lucrative and outrageous commerce in patent medicines. Morphine fell out of medical favor, as its newly recognized addictive properties seemed to rule out its use in many cases. To doctor and patient alike, the rise of addiction to heroin (a simple chemical modification of morphine) as a social problem made medical use of morphine increasingly objectionable. For many years, morphine and its close chemical relatives found use in medicine chiefly for the relief of severe pain in the terminally ill, where it was thought that opiate addiction could be condoned.

Today the attitude toward morphine is changing once again. After years of negative associations and rejection by physicians, morphine may be making a comeback as a legitimate medication. A 1980 study showed that when morphine was prescribed to relieve real pain in a hospital setting, practically none of the patients (4 out of 11,882) became addicted to it. Equally unforeseen, persons suffering from chronic pain can return to normal lives when their pain is controlled by morphine or related narcotics. Unlike heroin addicts on the street, these patients experience no euphoria or lethargy. After a brief period of adjustment to their medication, they are able to work, care for children, and drive an automobile. Earlier ideas about inevitable and incapacitating opiate addiction now seem to have been too simple.

Reports on managing chronic pain with morphine have been available for more than a decade, but clinicians have not rushed to prescribe opiates for suffering patients. Despite the favorable test results, a cruel circumstance diminishes the benefits that opiates can offer. This is not an insoluble biomedical problem but rather a deplorable social reponse. Patients who depend on morphine or other narcotics are often the victims of social stigma and discrimination. Employers, family members, and even physicians may look on these patients as drug addicts and treat them as social outcasts. Some patients who require morphine for relief from incapacitating chronic pain have been driven to extremes to hide their medical condition. One woman with a debilitating nerve condition has returned to a vigorous life on large doses of narcotics that she must take every three hours. In describing the stigma she faces, she says, "In a sense it would be easier to have cancer."

New developments in biotechnology may soon make long-term maintenance on narcotics easier. Investigators are now testing a tiny computerized morphine pump that can be implanted in a patient's abdomen for continuous pain relief. Because it delivers morphine directly into the spinal column, the effective dose is only one hundredth as large as that required for conventional administration. Another possibility is a patient-controlled pump that delivers pain relief on demand. If the burden of shame can be eliminated, the world's oldest treatment for severe pain may regain some of its former medical importance.

MALARIA AND THE PERUVIAN FEVER TREE

Morphine owes its medical renown to its efficacy in relieving pain rather than curing a disease. Before the advent of antibiotics, drugs that could cure diseases were rare, and infection was a very common cause of death. The first chemical that effectively controlled an infectious disease appeared in Western medicine around 1630 when the Spanish conquerors of Peru discovered that a local tree bark was useful in overcoming certain fevers. They brought the bark of this "fever tree," as they called it, back to Europe, and its fame quickly spread across the continent. The bark's active principle was quinine, and it soon proved to be a specific remedy for one of the world's most debilitating diseases. For nearly three hundred years quinine was the only effective treatment for malaria.

According to one delightful legend, the credit for introducing quinine to Europe belongs to Francisca, countess of Chinchón, who was the consort of Don Luis Gerónimo Cabrera de Bobadilla, count of Chinchón and viceroy of Peru. As the story goes, the countess lay in her palace in Lima languishing of a fever, and a band of faithful Indians (or in some versions, a wise Spaniard) came forward with an infusion prepared from the powdered bark of a local tree. The countess drank the medicine and rose from her bed, cured of her fever. In gratitude for her deliverance, she made the miraculous powder from the fever tree available to all who might need it. Jesuit priests bore it back to Rome, Spain, and all of Europe.

This romantic tale, or some variation of it, was widely accepted until the 1940s. Sadly, there is no historical evidence to substantiate the countess's role in bringing quinine to Europe, although she was indeed resident in Lima at the appropriate time. Long before the story was discredited, however, the great eighteenth-century systematist Carolus Linnaeus (Carl von

84

Linné) had given the name *Cinchona* to the fever tree in recognition of the countess of Chinchón and her philanthropy. (The first "h" in "Chinchón" was lost owing to a spelling error by Linnaeus.) Botanists now recognize several closely related cinchona trees (*Cinchona* species) whose bark contains varying amounts of quinine.

The historical fact best established concerning the arrival of quinine in Europe seems to be that by about 1650 the "fever-bark of Lima" was in use at the Hospital of the Holy Spirit in Rome. Over the next several years its use spread throughout Europe, and by 1677 it appears in the *London Pharmacopoeia* as *Tinctura Corticis Peruviani* ("tincture of Peruvian bark").

Although the Spanish brought cinchona bark from Peru to Europe, it is unlikely that they discovered its medicinal properties by themselves. Native civilizations had flourished in Peru for a thousand years before the Spanish arrived, and the Indians were excellent naturalists with a sophisticated pharmacopoeia of their own. Early Spanish chroniclers wrote admiringly of the Indians' knowledge of medicinal plants, and Spaniards had exploited this knowledge for years before the count of Chinchón arrived as viceroy in Lima. Within sixty years of the conquest in 1533, a steady stream of ships was carrying herbs, barks, and resins back to Spain for distribution throughout Europe. Already in 1568, a Spanish physician had written an account of new medicinal plants from Peru. Direct evidence is lacking, but local Indians probably revealed the secret of cinchona bark to their Spanish overlords. If so, they deserve credit for the first step in bringing quinine to the wider world.

However cinchona bark and quinine reached the world beyond Peru, they quickly became medically important as the only remedy for malaria. The powdered bark itself was used until the nineteenth century, when in 1820 two professors at the School of Pharmacy in Paris, Pierre Joseph Pelletier and Joseph Caventou, isolated quinine from cinchona bark. This, along with a second chemically related constituent, was the active drug. After Pelletier and Caventou's pioneering work, cinchona bark could be assayed for its quinine content and physicians could prescribe the drug more accurately.

The demand for quinine grew through the eighteenth and nineteenth centuries, particularly with European colonization of vast portions of Asia and Africa, where malaria was well established. In the nineteenth century, the growing consumption of quinine strained the supply. British India alone required 9 tons of quinine in 1850! The British began cinchona plantations in southern India, but the worldwide shortage remained unabated. Peruvian cinchona trees were recklessly overharvested to meet the expanding need, and finally the natural source of quinine was effectively exhausted. After some years of experimentation with cinchona seeds obtained from Peru, Dutch planters successfully established cinchona trees in Java. By 1872 these trees were mature enough to harvest, and for decades thereafter the Dutch enjoyed a monopoly in the world market for quinine.

Malaria, the disease responsible for this market, is an age-old scourge of mankind and one of the most serious human afflictions. As early as 2500 B.C., the Chinese were acquainted with an illness characterized by fever and enlargement of the spleen, symptoms characteristic of malaria. Although this disease was described very long ago, it may actually be quite recent in the history of human malaria. Most scientists believe that our primate ancestors suffered from an earlier type of malaria and simply brought the disease along as they developed into man. If this is correct, malaria has always been with us.

Despite the long human experience with malaria, the true nature of the disease remained obscure until recently. In the fifth century B.C., Hippocrates had pointed out that malaria was particularly common near swamps and bogs, and for centuries thereafter people assumed the disease was borne on noxious, swampy vapors that contaminated the air and infected those who breathed it. (*Mal'aria* means "bad air" in Italian.) Today it is common knowledge that malaria is a parasitic disease carried by mosquitoes that breed efficiently in swamps and bogs, but all this became clear only at the end of the nineteenth century.

We now know that four closely related species of relatively complex parasitic microbes (*Plasmodium* species) are the pathogens that cause malaria, and that each of these four species leads to a somewhat different form of the disease. These parasites are transmitted to humans by the females of several species of *Anopheles* mosquitoes. Like the nematode responsible for onchocerciasis, *Plasmodium* has a complex life cycle and passes through multiple forms in both insect and man. One form invades its human host's red blood cells, where it multiplies, eventually causing the red cells to rupture. When a sufficiently large number of red cells are attacked, the human host begins to suffer the cycles of chills, fever, and sweating that characterize malaria.

Most of us living in developed countries never directly experience malaria, and its importance in human affairs may be difficult to appreciate. Quite simply, malaria is the world's worst health problem. In its various forms, it is both a swift killer and a chronic, debilitating sickness. WHO estimated a few years ago that well over 2 billion people are exposed to malaria, and that 270 million are infected. More people have malaria than any other disease, and 1 million or more of them die each year. Most of those who die are children in the tropics. There is also increasing evidence that global warming will spread malaria and other tropical diseases into temperate regions previously not at risk.

A disease of this scope and severity can influence the course of history, and malaria has done so repeatedly. It has been responsible for social, economic, and political upheavals from classical times to our own era. An outbreak of malaria was a crucial factor in the total destruction of the Athenian army at Syracuse in 413 B.C., and this defeat was a disaster from which Athens never recovered. More recently, malaria helped vanquish McClellan's Army of the Potomac in the Chickahominy Valley of Virginia in 1862, and it sped the American and Filipino surrender of Bataan and Corregidor to the Japanese in 1942. Less dramatic perhaps, but no less deadly, are the civilian epidemics of malaria that strike from time to time. The worst occurrence in the past century was in 1923 in the then Soviet Union, where 5 million cases of malaria led to over sixty thousand deaths.

The importance of quinine in fighting this dreadful disease would be hard to overestimate. Until about 1930, it was the only effective medication. If malaria has influenced the course of history, quinine has doubtless done so as well. Without quinine to ward off the ever-threatening fevers, European colonization of tropical Africa, Asia, and the New World would have been a much different story. Whether they went as missionaries, colonists, or fortune-seekers, Europeans (and later, Americans) could have never survived malaria in their distant outposts without the protection afforded by quinine.

By the early twentieth century, a stable and profitable industry fulfilled the world's need for quinine. In the 1930s, however, the situation began to change. Synthetic antimalarials appeared in Germany and began to find acceptance. The initial trend toward synthetic medications was assured during the Second World War. Much of the fighting happened in malarious regions of the South Pacific, Asia, and Africa, where the Allied forces required protection against the disease. But quinine had disappeared from the world market early in the war when Java and its cinchona plantations fell to the Japanese. Replanting cinchona trees elsewhere would have been futile, because new trees require years to become productive. Military medical officers urgently sought a new antimalarial. The resulting program was second in priority only to the development of penicillin in wartime drug research in the United States and Britain. Chemists prepared thousands of compounds for biologists to test. The quest eventually led to a very effective synthetic compound that had some chemical similarity to quinine and to earlier synthetic antimalarials. The scientists named the new compound chloroquine.

After the Second World War, final conquest of malaria appeared within reach. Chloroquine and related synthetic drugs were stopping the disease, and the new chemical pesticide DDT was effectively eradicating *Anopheles* mosquitoes. Public health officials were optimistic until the 1960s, when incipient resistance to both chloroquine and DDT began to darken their outlook. Falciparum malaria in South America (caused by *Plasmodium falciparum*), the most extreme, often fatal, form of the disease, was no longer responding to treatment with chloroquine. A new mutant strain of the parasite had appeared, and it was resistant to the drug. This resistant strain took hold and spread inexorably around the world.

Malaria, so recently in full retreat, began an appalling comeback. WHO campaigns in the 1950s had brought the number of cases of malaria in India down to fifty thousand by 1961. As resistant *Plasmodium falciparum* penetrated Asia, this number began to rise. By 1969 there were 350,000 cases in India, and the number doubled in 1970 and again in 1971. In 1977 some estimates suggested at least 30 million cases of malaria in India! Most seriously, the proportion of falciparum malaria cases greatly increased.

When chloroquine proved no longer effective, public health officials were forced to fall back on quinine as the only other treatment for falciparum malaria. There was no question that quinine was effective, but it had been off the market for twenty years. Chloroquine had replaced quinine as the treatment of choice, and after the Second World War the cinchona planters of Java had turned to other crops. There was no longer any natural source of a medication once shipped around the world by the ton. Locating a supply of quinine meant combing through drug warehouses for dusty bottles with crumbling labels. Eventually, renewed demand for quinine convinced growers to reestablish cinchona plantations, but it took some years for these new plantations to become productive.

In recent years new antimalarials have appeared, but quinine remains in use to treat falciparum malaria. Travellers to regions where resistant strains of *Plasmodium* hold sway still receive quinine as a preventative. For the present at least, quinine has recaptured a meaningful role in combatting the world's worst disease.

FROM WILLOW BARK TO A SUCCESSFUL DRUG

Morphine and quinine are only two of thousands of medicines extracted from plants over the centuries. Historical connections between plants and medicines are complicated, and they are not always obvious. Well-known medicines sometimes have plant origins that have since disappeared from historical sight, and some plants familiar to us were the sources of medicines once popular but now no longer in use.

Probably the best-known and most widely used medicine associated with plants is aspirin. Although aspirin does not come from a plant, it was specifically designed to compete with several popular plant medicines. The active ingredients in these medicines all had similar chemical structures and provided the inspiration for aspirin. The natural medicines were in wide use throughout the nineteenth century, but aspirin largely replaced them in the early 1900s.

Unlike quinine, the plant products that inspired aspirin have a well-documented historical origin. They began with a discovery by an eighteenth-century English clergyman, the Reverend Edward Stone, who lived not far from Oxford in the village of Chipping Norton. In 1757, according to his later report, Reverend Stone accidentally tasted the bark of a white willow tree (*Salix alba*). As he wrote to the Royal Society to announce his discovery, "I was surprised by its extraordinary bitterness, which immediately raised me [sic] a suspicion of its having the properties of the Peruvian bark." The Peruvian bark was well known for treating fevers, and Reverend Stone suspected that willow bark might also be a fever remedy. He considered his reasoning all the more convincing because willows grow in wetlands and wetlands are associated with fevers. He cited "the general maxim, that many natural maladies carry their cures along with them, or that their remedies lie not far from their causes."

Guided by these considerations, Reverend Stone tested the medicinal properties of willow bark. He carefully dried the bark, powdered it, and used the powder to treat persons suffering from fevers. He treated about fifty cases over five years and concluded that willow bark was a "very efficacious" remedy for fevers. Reverend Stone's reasoning was faulty, but it had led him to an impressive discovery.

The active principle in Reverend Stone's willow bark was salicin, a derivative of a chemical known as salicylic acid. Over the next eighty years, drugs from two other plants, meadowsweet (*Filipendula ulmaria*) and wintergreen (*Gaultheria procumbens*), also yielded active principles chemically similar to salicylic acid. The experience of another fifty years showed that these salicylates, as they are called, effectively reduced fever, pain, and swelling, and relieved gout, rheumatic fever, and arthritis. Such favorable medicinal properties led the chemical research laboratories of the Bayer company in Elberfeld, Germany, to seek a new salicylate for the drug market. The goal was to prepare a synthetic compound that could compete successfully with the natural salicylates then available. Nineteenth-century chemical knowledge and technology were adequate to the task, and Bayer's research led in 1897 to a compound with the chemical name *O*-acetylsalicylic acid, which we now know as aspirin.

Aspirin has had an extraordinarily successful history. Owing to its low toxicity and limited side effects, it has long been available to the public without prescription. The result is a worldwide consumption of about 100 million pounds (over 45,000 metric tons) of aspirin a year. Americans buy over 30 billion aspirin tablets annually. Most often people take aspirin for relief from simple headaches and other minor pains, but aspirin is also an important drug in treating rheumatoid arthritis and in preventing blood clots by blocking the aggregation of blood platelets. An exciting finding in recent years is that a regular low dose of aspirin diminishes the incidence of heart attack, probably owing to inhibited platelet aggregation. Another boost for aspirin came in mid-1993, when the American Cancer Society released a study suggesting that people who take aspirin regularly have a 40 percent lower death rate from cancers of the digestive tract.

Aspirin continues to find new uses, but over the years many other plant-derived drugs have come and gone. Some once popular plant medicines have disappeared or changed in use with the passage of time. Tobacco had a long history as a religious and medical drug in the New World before the Europeans arrived, and it first reached Europe as an ornamental and medicinal plant. Sixteenth- and seventeenth-century European physicians prescribed tobacco for a variety of diseases, including asthma and cholera, and sometimes administered a tobacco infusion by enema. Medicinal use of tobacco declined in Europe as more effective drugs became available, although it is still widespread among South American Indians.

Another medicinal plant that has fallen into disfavor is garlic (*Allium sativum*), but garlic failed for different reasons, largely social ones, despite strong evidence of its virtues and a long history of medical use. The Egyptians of the early New Kingdom (about 1500 B.C.) included garlic as a key ingredient in several medications. A thousand years later, Hippocrates recommended garlic to the Greeks. Imperial Roman soldiers ate garlic to rid themselves of intestinal worms, and sailors as far separated in time as the ancient Phoenicians and medieval Vikings regularly carried garlic as a remedy for illnesses at sea.

All these peoples must have regarded garlic's strong taste and odor as evidence of its potency. Garlic deserved its reputation as an effective medication, because it does have a number of useful properties. Garlic juice inhibits the growth of many pathogenic microorganisms, including those that cause dysentary, cholera, and human yeast infections. It retards the aggregation of blood platelets, lowers blood-cholesterol levels, and enhances the assimilation of vitamin B_1. Recently, garlic has been found to inhibit both lung and colon cancer in experimental animals.

Garlic preparations are sold in many American drugstores, but they probably find more medical acceptance in Europe than in the United States. Nonetheless, garlic has failed to win many supporters in mainstream medicine. Some of the active principles are unstable, and good alternative

91

medications are available. Besides, many people associate garlic with an odor they find repellent.

SEEKING A DIFFERENT REALITY

Medicines from plants are the largest group of useful preparations that humans have derived from nature, but there is a much smaller group of botanical agents that probably excite more interest. These are the natural mind-altering, or psychotropic, drugs. In general, these drugs act either to mimic or to block the action of a chemical compound that transports nerve signals in the brain. Some of these drugs cause hallucinations; others induce psychosis or euphoria. Still others act as sedatives, hypnotics, or stimulants.

Natural mind-altering drugs come mostly from plants or fungi. For several thousand years, they have inspired awe and reverence in peoples around the world, offering a miraculous escape from the everyday world into a spiritual realm. Because they appear to open the way into another world, psychotropic agents have held a prominent position in both religion and medicine. Many traditional cultures attribute sickness and death to malign spirits, so that medicine and religion become inseparable. For priests or healers to diagnose and treat diseases, they must conciliate the appropriate spirits. Under the influence of psychotropic drugs, a priest can enter the world of these spirits, communicate with them, and then work to heal the patient. (From this point of view, psychotropic agents should not be separated from medicines, as we have done here.) Some traditional cultures limit access to mind-altering drugs to priests and other special members of society. Others make these drugs available to ordinary people as well.

Psychotropic drugs have turned up in cultures in all parts of the world. We have seen that several early Mediterranean cultures used opium both to kill pain and to induce euphoria, and that opium served both religious and medical needs. Other mind-altering drugs appear in Old World civilizations, but they are much more common in the Western Hemisphere. It is not clear why this geographical bias exists, because there is no botanical reason to think that psychotropic drugs are more prevalent in New World plants.

What is clear is that, for at least fifteen hundred years, native peoples of the Western Hemisphere have used local plants to prepare a remarkable variety of concoctions to eat, drink, smoke, or chew. When Columbus landed at Hispaniola on his second voyage in 1496, he found the Taino Indians using something they called *cohoba* to communicate with the spirit world. Cohoba is a powerful hallucinogenic snuff made by drying and grinding large beans that the Indians harvested from the yopo tree (*Anadenanthera peregrina*). A little later in Mexico, the Spanish invaders discovered ololiuqui, an Aztec hallucinogenic drink made from the seeds of the snakeplant (*Turbina corymbosa*). Hundreds of years later, scientists found that snakeplant seeds contain compounds related to LSD, a synthetic hallucinogen that was not prepared until the mid-twentieth century.

Farther south, the conquistadores encountered a milder psychotropic agent. The Incas in Peru chewed the leaves of the coca plant (*Erythroxylon coca*) as a stimulant and analgesic. The habit was not widespread among the working people, because coca leaves were also a form of currency. However, the Spaniards were soon urging coca on the Incas they forced into service as gold miners. They had discovered that coca was an ideal drug for impressed workers: it suppresses hunger and sustains heavy manual labor.

Coca was in use in Peru a thousand years before the Spanish arrived, and the Incas' modern descendents continue to chew coca leaves, currently at the rate of about 13 million pounds a year. Perhaps more meaningful for us, the active principle in coca is cocaine. Clandestine drug laboratories in northern South America process tons of coca leaves to supply cocaine as a recreational drug to an evidently insatiable American market. Cocaine and the quick money it promises have brought drug gangs, chaos, and murder to cities and towns across the United States.

Much of what we know about New World psychotropic drugs we owe to Richard Evans Schultes, an ethnobotanist at Harvard University who has spent a lifetime studying hallucinogenic drugs, the plants they come from, and how Indians used and still use them. Schultes's doctoral thesis in 1937 was a guide to peyote, one of the best-known hallucinogens. Ritual use of peyote was established long before the Spanish arrived in Mexico, and it is still practiced by a number of Indian tribes there. In addition, peyote is legally protected in the United States for use in the ritual of the Native American Church, an institution that claims 250,000 members. In 1993, the

United States Congress reconfirmed this traditional use of peyote as a matter of religious freedom.

The drug at the center of these activities comes from the peyote plant, a small, round, gray-green cactus (*Lophophora williamsii*) that grows from the Rio Grande Valley in Texas south to north-central Mexico. In several Amerindian languages, the word for "peyote" is the same as the word for "medicine." The main active principle in peyote is a chemical compound called mescaline, which is concentrated in the upper portion of the cactus. To harvest peyote, the round tops of the growing cactus are cut off parallel to the ground and then dried. These dried discs, or mescal buttons, are tough and essentially indestructible. A peyote user simply takes a button into his or her mouth, moistens it with saliva, and swallows it. In earlier times many Indian tribes made pilgrimages to gather peyote in areas where it grew. Now, members of the Native American Church as far north as Canada mail-order supplies of mescal buttons from Texas.

A second widely known vision-inducing drug from Mexico is psilocybin, the active principle of mushrooms that the Aztecs considered sacred and called *teonanacatl,* meaning "divine flesh." We know these fungi as *Psilocybe mexicana* and several related species. The Aztecs ate these mushrooms to commune with the spirit world, which in turn permitted them to treat the sick and to engage in divination and prophecy. Mexican frescoes and Guatemalan stone sculptures suggest that the mushroom cult was thriving by A.D. 300. Today, *Psilocybe* species are widespread in North and Central America and some modern Indian tribes use them much as the Aztecs once did.

Our knowledge of the modern use of *Psilocybe* originates with the field work of an enthusiastic and persistent amateur. Robert Gordon Wasson was an American business executive with a passionate interest in mushrooms. He gained the confidence of the Mazatec Indians in southern Mexico, and in June 1955, Wasson and a companion became the first outsiders to consume the sacred mushrooms under the guidance of an Indian healer.

Among the Mazatecs, taking the mushrooms was a private activity rather than a tribal rite. Wasson and his companion joined a group of several local people, and they all ate fresh mushrooms in a simple ritual led by a local woman and her daughter. After consuming his mushrooms, Wasson lay on a mat for several hours. He was conscious and aware of his local

5.2 Archeologists have excavated great numbers of these "mushroom stones" from highland Mayan sites in Guatemala, some from as early as 500 B.C. The stones were probably icons connected with mushroom worship.

surroundings. At the same time, he also had vivid hallucinations of dramatic landscapes. His later description of the scenes he beheld hints at the power of psilocybin:

> [The scenes] were of a vast desert seen from afar, with lofty mountains beyond, terrace above terrace. Camel caravans were making their way across the mountain slopes . . .
>
> [T]he landscapes responded to the command of the beholder: when a detail interested him, the landscape approached with the speed of light and the detail was made manifest. When I was seeing the camel caravans in the distance, an impulse seized me: right away I was upon them, listening to the

sounds of their heavy breathing, to the camel bells, to their lurching with their loads, smelling their stench.

MARIHUANA AND CANNABIS

Before leaving our discussion of psychotropic drugs, we must say something about marihuana, which became notorious as a recreational drug in the United States in the 1960s and 1970s. In the social and political unheavals of those years, marihuana was a sign of rebellion and the subject of fierce public debate. Defiant students grew pot in window boxes, earnest academics wrote proposals for marihuana research, and concerned legislators drafted bills and appointed committees. The National Commission on Marihuana and Drug Abuse estimated in 1972 that 24 million Americans, about 15 percent of the population over eleven years of age, had tried marihuana at least once. The percentage among well-to-do high school and college students was much greater.

Long before this national uproar, the United States government had moved to criminalize the use of marihuana. Around 1910, Mexican-Americans in the Southwest had introduced marihuana into the United States as a recreational drug, but for several decades its use was confined to poor and marginalized groups in society. The federal government responded with the Marihuana Tax Act of 1937, which sought to eradicate the drug from the United States. The Bureau of Narcotics warned Americans of marihuana's dangers in posters branding it as a "killer drug" and "a powerful narcotic in which lurks Murder! Insanity! Death!" Just what formed the factual basis of these dreadful warnings is not clear. Thirty years later, however, these earlier statements lent support to similar claims that pervaded some segments of American society. Marihuana became associated with student protest and symbolized one side of the national controversy over drugs, sex, authority, and the war in Vietnam. Warnings about its dire effects became routine. In the disarray of those unhappy times, it was easy to overlook the fact that millions of people had been using this drug in one form or another for over two thousand years.

The source of marihuana is the hemp plant (*Cannabis sativa*), a hardy weed that is a prehistoric source of fiber for cordage. The major psychotropic compound in hemp is tetrahydrocannabinol (THC), which is particularly concentrated in a resin associated with flowers of the female plant.

Cordage fiber comes from the male plant. Marihuana is a mixture of female flowering tops and leaves that have been dried and chopped or ground. It is usually smoked like tobacco. In other parts of the world people smoke, eat, and drink various other derivatives of hemp. Some of these concoctions are considerably stronger than the marihuana ordinarily available on American streets. For convenience we can refer collectively to all preparations containing THC as cannabis. They have a long history of religious and medical use, and they have also been available to common people for centuries.

Cannabis is the most widely disseminated hallucinogen in the world today, and hemp is one of the world's oldest cultivated plants. It was a source of fiber in northeastern China by about 4000 B.C., and there is some evidence that the Chinese were using cannabis as a sedative as early as 2700 B.C. The hemp plant thrives in warm climates, and sometime before 800 B.C., it spread to India. There, holy men adopted it for religious rites, and today hemp is still found in the gardens of many Indian temples. In India, cannabis also became a medicine prescribed for a variety of ills, and at some early date it passed into the hands of ordinary people.

Cannabis soon diffused into Persia. Details of its progress farther westward for the next several centuries are uncertain or disputed. A recent discovery near Jerusalem establishes that cannabis had reached the eastern Mediterranean by roughly A.D. 300. In June 1993, Israeli archeologists announced the excavation of a fourth-century tomb that contained the remains of a teenage girl carrying a full-term fetus. She had apparently died in childbirth or near the end of her pregnancy. The archeologists found *Cannabis sativa* ashes near the skeletal remains and suggested that the plant had been burned and the smoke used as an inhalant to ease the pain of childbirth.

Several hundred years later, the rise of Islam facilitated diffusion of cannabis along the southern shore of the Mediterranean. The Koran forbids indulgence in alcohol but not cannabis, and the Muslim Arabs embraced the drug. When they began their sweep across North Africa in the ninth century, they took cannabis with them. They greatly extended its range but failed to establish it in Europe. Spain was the only land under Arab conquest that resisted the new drug.

Cannabis came to America as a drug only in the nineteenth century, although American settlers cultivated hemp earlier for its fiber, and the plant had first arrived in the New World with the Spanish conquistadores. Much like opium, cannabis became an all-purpose remedy and a popular ingredient in patent medicines. Tilden's Extract of Cannabis Indica was widely available during these years, and we may suppose that it was a soothing and satisfying medication. Despite this easy accessibility, Americans showed little interest in cannabis as a recreational drug until the early twentieth century.

Marihuana today is still readily available on the street and around college campuses in the United States. About 28 percent of American students used marihuana in 1992, and use among high school seniors increased sharply in 1993 after years of decline. The increase continued in 1994, and drug experts are concerned because they say marihuana is now much more powerful than the marihuana of the 1960s and that use is no longer tied to political protest. Nonetheless, the excitement seems to have faded. Alcohol returned as American students' drug of choice in the 1980s, and if there are those who prefer marihuana, few seem to notice or care much now. Perhaps the most significant legacy of the marihuana controversy of the sixties and seventies is the research it engendered. We learned more about *Cannabis sativa* and THC in a decade than in all the preceding centuries that cannabis has been in use.

DISCOVERING PLANT MEDICINES: WHO WAS FIRST?

People have been nibbling at unfamiliar plants for ages, looking for new foods and for new sources of medicines and other useful preparations. Someone, or maybe several people, somewhere in South America, must have first noted the salubrious effects of cinchona bark and passed this information on to others in the community. Each of thousands of plant medicines discovered over thousands of years has its history, but early peoples seldom left records for posterity. We cannot hope to unearth the detailed history of most of their findings. We often have a discovery but know nothing about the discoverer.

There is another, perhaps less obvious uncertainty associated with the origins of plant medicines. Humans are not the only herbivores that regularly investigate plants for new foods. Many other species do the same. Is it possible that some of these other species also discover medicines and pass this knowledge on in their own communities? Do other species use plants to fight diseases the way humans do? If so, the history of plant medicines could be older than the history of mankind.

Biologists have long discussed circumstantial and anecdotal evidence that other species use medicines. There is, for example, a plant in India known as pigweed (*Boerhavia diffusa*), from which local people extract an antihelmenthic, or worm-fighting, agent. Wild boars (*Sus scrofa*) dig up pigweed and eat the roots with unbridled enthusiasm. On the other side of the world, Mexican pigs (also *Sus scrofa*) are said to be very fond of pomegranate roots (*Punicum granatum*), which also contain an antihelmenthic agent. Pigs everywhere typically support a large population of parasitic worms, so it is tempting to conclude that Indian and Mexican pigs eat antihelmenthic roots to improve their health. The conclusion may seem farfetched, but we saw earlier that starlings control insects and other parasites by adding fresh green matter to their nests. If animals control external medical problems with plants, why not internal ones? The conclusion is plausible, but the only certainty is that these observations about pigs and roots are difficult to interpret.

Because anecdotal information is questionable, reports from professional animal behaviorists about herbivores and medicines have aroused considerable attention. One of the most provocative of these is a report by two investigators of primate behavior, Richard Wrangham and Toshisada Nishida, then at the University of Michigan and the University of Tokyo, respectively. In 1983 at Tanzania's Gombe and Mahale Mountains National Parks, Wrangham and Nishida watched the peculiar way chimpanzees (*Pan troglodytes schweinfurthii*) select and eat the young leaves of certain herbaceous plants (*Aspilia rudis* and related species).

The chimpanzees select each leaf with elaborate care, sometimes closing their lips over a leaf and holding it for a few seconds before deciding whether to pluck it from its branch. They place each chosen leaf in their mouth, roll it around, and rub it with their tongue against the inside surface of their cheek. Then they swallow the leaf without chewing it, even though *Aspilia* leaves are covered with bristly hairs and must be difficult to swallow whole. Chimpanzees do chew other leaves that they pick and eat. The animals seem to consider the whole *Aspilia* experience unpleasant. The behaviorists saw one chimpanzee vomit while feeding on the plant, although afterward he seemed healthy and vigorous for the rest of the day. Another chimpanzee wrinkled his nose repeatedly as he swallowed the leaves. Wrangham ate a leaf himself and found it "extremely nasty."

Were these chimpanzees taking medicine? Several observations strengthen the possibility. Chimpanzees cannot gain much nourishment from *Aspilia*. They never chew the leaves, and leaves retrieved from feces appear fresh, perhaps folded once or twice but otherwise hardly damaged in their passage through the digestive tract. The indigenous Africans value *Aspilia* as a medicinal plant. They use the leaves and roots in treating a number of topical ailments, such as burns and wounds, as well as for stomach pains and worms.

The chimpanzees' curious habit of not chewing the leaves is reminiscent of the preferred mode of administering a few drugs in human medicine. The best known of these is nitroglycerine, which acts directly on the heart to alleviate the chest pains of angina pectoris. The patient holds a pill of nitroglycerine under his tongue, and the drug is absorbed directly into the blood. In this way nitroglycerine enters the circulation more rapidly than if swallowed. It can reach the heart without destruction by liver enzymes or by stomach acid. In the same way, perhaps *Aspilia* leaves are a more effective medicine if they are not chewed. This is a reasonable possibility, because the leaves contain thiarubrine-A, which is an antibiotic that is destroyed by acids. Chimpanzees may get more thiarubrine-A into their circulation by direct absorption through the mouth than they would from leaves that were chewed, swallowed, and exposed to stomach acid.

Several other genera of plants in the same family (Asteraceae) as *Aspilia* also produce thiarubrine-A. In fact, one of these plants is important in the traditional medicine of a group of Indians in the Pacific Northwest. Thiarubrine-A is a potent antibiotic, antihelmenthic agent, and fungicide. It is effective against pathogens in the dark but even more toxic in the light, indicating that it is both poisonous and phototoxic.

Other observers have reported similar leaf-swallowing behavior by baboons (*Papio anubis* and *Papio hamadryas*), and there are also descriptions of chimpanzees eating other species of medicinal plants. These observations reinforce the argument that these primates are indeed taking medicine. A particularly striking report came from Michael A. Huffman of Kyoto University and Mohamedi Seifu of the Mahale Mountains Wildlife Research Centre in 1989. Huffman and Seifu followed a female chimpanzee and her two-and-a-half-year old infant closely for two days in Mahale Mountains

National Park in Tanzania. The mother chimpanzee, whom the behaviorists called CH, appeared to be ill on the first day they observed her. She was lethargic, rested much more than usual, and left her baby CP in the care of other chimpanzee mothers.

That first afternoon CH was sluggish, trailing slowly after two other mothers with their three offspring and CP. When the group stopped to feed on a common grass, CH ignored the grass and went directly to a 2-meter-high shrub of *Vernonia amygdalina*. She pulled down several young branches and peeled off the outer bark to expose the pith. She bit a stalk into short portions and then chewed each portion for several seconds, making sucking sounds as she extracted the juice from the pith. CH swallowed the juice and spit out the fibrous pith. She briefly joined the other chimpanzees and then returned for more shoots from the shrub. Another mother, WD, was nearby eating a grass stalk, which WD ate while sitting on top of a bent over *V. amygdalina* shrub, but she paid no attention to it. CP became interested in what his mother was eating. He repeatedly begged CH to give him some of the shoots from her mouth. Picking up pieces of stalk that CH dropped, he put them in his mouth and quickly discarded them. He then returned to feeding with WD. That afternoon, CH appeared to have bowel trouble, and her urine was dark colored.

The next day CH started out slowly. She lay down at the edge of the group, leaving CP to play with a young adult and two adolescents. Within two hours, however, her appetite and stamina seemed to return. That afternoon, CH rejoined the group; she fed and behaved normally thereafter.

The gist of Huffman and Seifu's observations is of course that CH appeared to be ill, she fed on a plant, and then she got well. As with the report on *Aspilia*, other facts reinforce the idea that CH was taking medicine. Like *Aspilia*, *V. amygdalina* is a well-known medicinal plant. It grows widely throughout tropical Africa, and all across this broad area local people use it to treat parasitic diseases, intestinal upsets, and other conditions. The plant is called bitterleaf in English, and its leaves and stems contain a toxic bitter principle. *V. amygdalina* is in the chimpanzees' natural diet, but they feed on it infrequently despite its availability throughout the year. In general, chimpanzees avoid feeding on plants that taste bitter to humans, so *V. amygdalina* is a conspicuous exception to this rule.

These reports from Tanzania and similar studies led to the first symposium to discuss research in zoopharmacognosy, as this area of investigation has been named. In early 1992 the American Association for the Advancement of Science (AAAS) sponsored a meeting at which animal behaviorists discussed suggestive evidence that not only primates, but also bears, dogs, and cats use medicines. Among the presentations were new studies on the Tanzanian chimpanzees. It now appears that chimpanzees feed more on medicinal plants in the rainy season than during the rest of the year. Since the rainy season brings more disease, this behavior might indicate that chimpanzees practice preventive medicine. Another presentation suggested that those chimpanzees most infested with intestinal parasites feed most frequently on medicinal plants. It seems increasingly likely that chimpanzees do indeed use medicines. There should be more information on dogs and cats, bears and pigs in the coming years.

WHAT ABOUT THE FUTURE?

If other living species do in fact exploit plants to improve their health, mankind may have been using plant medicines since its earliest days. But what about the future? Will plants continue to be an important source of human medicines? Nowadays, biomedical scientists uncover the details of one disease after another. Clinical medicine has available new magic bullets each year. In this high-tech world, plant medicines may seem a bit quaint. History and romance apart, do we really care any longer about barks and berries with therapeutic properties?

The short answer is that we still need plants. About a quarter of the drugs in prescription medicines today are derived from plants. A survey in the mid-1980s indicated that American consumers spend about $8 billion a year on plant-derived prescription drugs. According to more recent figures, the twenty top-selling drugs in the United States have annual sales of $700 million to $2 billion each, and six of these drugs come from plants.

Pharmaceutical companies have recently increased their investment in the search for new medicinal plants. One reason for this is the prevalence of diseases and conditions that we cannot yet combat with specifically designed drugs. There are still no magic bullets for AIDS or cancer. There is no pill for treating stroke. Plant products sometimes promise the best available treatment for such conditions.

Why should plants produce compounds that are effective against AIDS or stroke, when these are threats that plants never face? Why should defenses evolved over innumerable generations to oppose a plant's enemies sometimes be effective against quite different human afflictions? It is not obvious why this is so, but a fascinating possibility is that, on a molecular level, nature's bag of chemical tricks may be limited. Perhaps plant and human predators sometimes employ closely related means of attack that can be blocked by the same defense.

This suggestion may seem unreasonable, but it is related to recent discoveries about how evolution operates. The emerging idea is that evolutionary development proceeds by tinkering with what is at hand rather than by launching grand new enterprises. Biologists are beginning to recognize that organisms develop new structures and capabilities by adapting already existing molecules to new uses. An evolutionary development may look quite novel, but close similarities between the new and the old may emerge on examination at the molecular level. If this idea is correct—if nature is generally conservative in fashioning novel capabilities—predators may have a limited number of weapons in their arsenals. Regardless of whether their victims are plants or humans or any of thousands of other creatures, there may be only a limited number of ways they can conduct their assaults.

It is also possible that plant products are particularly useful drugs for reasons related to acquired resistance. We have seen that organisms respond to toxic compounds in their environment by developing resistant strains. The diamondback moth has acquired resistance to every pesticide available, and we spend $1 billion a year around the globe trying to control it. In the same way, pathogens that we wish to destroy may become resistant to our synthetic drugs.

There is some evidence that pathogens generally develop resistance more readily to laboratory chemicals than to plant compounds that have survived natural selection for countless generations, even when the pathogens are different from the natural targets of the plant compounds. Quinine is a good example of this idea. Quinine continues to be effective against *Plasmodium* species even after three hundred years of use, while *Plasmodium falciparum* acquired resistance to chloroquine, a synthetic compound, within twenty years. It is still far from certain that resistance is generally slow to develop against plant-derived drugs, but our experience with quinine and some other drugs indicates that this is an important concept for future investigation. If this advantage concerning resistance is real, we would do well to seek new drugs from nature.

Ethnobotanists, private pharaceutical houses, and the National Institutes of Health (NIH) have all realized the advantages of plant-derived medicines, and all are screening plants for new drugs. Two sorts of programs are underway. The first centers on exploring systems of traditional medicine. Many cultures make use of traditional medicinal plants about which we have no modern scientific information. A careful enumeration of traditional Chinese remedies in 1980 listed 2,270 medicines. Most of these represent discrete species of plants, many of which are readily available in local markets in China. A study of herbal medicine in one small region of northern Thailand turned up 553 species of local plants in current use. The NIH's National Cancer Institute is now examining medicinal plants from China and Korea, and the large American drug house of SmithKline Beecham is screening plants native to Ghana and Malaysia. One encouraging lead turned up by the NIH study is an anti-carcinogen from a Chinese source.

Ethnobotanist Paul Alan Cox of Brigham Young University is gathering information on plants from traditional Samoan healers. Eighty-six percent of these plants have shown pharmacological activity in broad screening tests, including a number of compounds with high levels of antiviral, anti-tumor, or anti-inflammatory activity. Cox's most interesting finding so far is a compound with promising activity against HIV, the virus responsible for AIDS. There is some sense of urgency in these studies of traditional healing, owing to the progressive disappearance of indigenous medical traditions. Tribal peoples are rapidly being acculturated and assimilated into the wider world, and the sources of many medicinal plants are vanishing before expanding populations and modern development.

In addition to plants used in traditional medicine, there are thousands of other species that have never been screened for possible biological activity. Biomedical scientists have examined thoroughly only about 0.5 percent of the world's higher plants for medically useful activity. The greatest concentration of unexamined species is in the tropics. (Of the known 250,000 species of higher plants, about 110,000 grow only in the New World tropics.) Several studies are now underway to explore this untapped source of natural compounds. One of the most interesting of these is a program sponsored by Merck & Company, the American pharmaceutical company we discussed earlier. In 1992, Merck agreed to pay the Costa Rican National Institute of Biodiversity $1 million over two years for the right to search for new drugs in the tropical forests of Costa Rica. Local residents, who were first trained in collecting and identifying species, are now gathering samples and cataloging plants and animals. Biologists estimate that more than five hundred thousand biological species are native to Costa Rica; fewer than eighty-five thousand of these have been scientifically described and catalogued.

An important feature of this agreement is that Merck and the National Institute of Biodiversity will share royalties from any successful drugs that emerge from the survey. Half these royalties, as well as 10 percent of the initial one million-dollar payment, are designated specifically for conservation work in Costa Rica. Conservationists have a strong interest in how this agreement works in practice, because it offers an attractive arrangement for saving tropical forests from further destruction. Cutting down a forest brings quick cash from the sale of timber and often provides land for a growing population of destitute peasants. These are immediate advantages that a developing nation finds hard to resist. To forego the cash and protect its forests, a poor country requires an alternative source of profit from its national resources. Harvesting drugs may offer a way to make forests profitable without destroying them. If the country of origin shares in the profits, drug harvesting could turn a tropical forest into a sustainable national resource. A forest could have a long-term value much greater than the one-time profit realized from clearing it for timber and development.

The government of Costa Rica is seriously committed to maintaining its biological resources and dedicating them to improving the quality of life. As the first official act after his inauguration in May 1994, President José María Figueres opened a symposium on Costa Rican biodiversity and its value to the country. Other nations that are poor in cash but rich in biological resources, including Mexico, Nepal, and Indonesia, are watching the Costa Rican experiment with interest.

The search for new medicines in their forests may give developing countries an economic rationale for saving these scientifically unexplored areas. Furthermore, it may offer the developed world a welcome reason to favor preservation. People who have been unimpressed by efforts to save forests as reservoirs of biological diversity and ecological balance for their own sake may be more interested in maintaining these areas as sources of novel medicines that will benefit both the developed and the developing worlds.

The tropics are not the only potential source of medicines. Our own native plant species also offer opportunities. In September 1993, the New York Botanical Garden and Pfizer, another of the country's large pharmaceutical houses, announced a joint venture to survey this domestic reservoir for new drugs. Pfizer will finance a $2 million program of the Botanical Garden to collect and identify plants from across the United States. If Pfizer scientists derive commercial drugs from these plants, the Botanical Garden will receive royalties.

In these surveys of uninvestigated species and traditional medicinal plants, there is always the possibility of finding a compound that is an immediately useful drug. It is more likely, however, that the plant chemicals discovered will serve as leads for drug development. Perhaps only minor modification will be necessary to convert a new compound into a good drug, as in the conversion of the avermectins from *Streptomyces avermitilis* into the commercially important ivermectin. In other cases, new compounds will simply suggest unprecedented types of chemical structures for synthesis in the laboratory, much as the natural salicylates inspired the synthesis of aspirin a century ago.

6

CHEMICALS AND
LIFESTYLES

WE NOW turn away from chemicals for warfare to examine a different group of nature's compounds that facilitate or sustain many organisms' particular way of life. Such compounds are the chemical equivalent of teeth for a shark or wings for a bird. When we mention sharks and birds, teeth and wings come to mind as structural adaptations that essentially define these creatures. In similar fashion, certain chemical compounds define chemical adaptations. These chemicals may be less obvious than teeth or wings, and they are certainly much less complex, yet they can be just as critical to a particular lifestyle.

There are chemicals that allow certain fish to thrive in the freezing oceans surrounding Antarctica. There are chemicals that enable oysters to house themselves in rugged shells or spiders to trap their prey in well-made webs. Still other compounds lend protective coloration to skin, hair, and feathers, or permit lightning bugs to make their own light. Lifestyles consist of many components, but specific "lifestyle-chemicals" are often necessary for their success. The compounds themselves are sometimes more chemically complex than most warfare agents, but like them, lifestyle chemicals have survived generations of evolutionary testing and refinement.

Chemical adaptations have attracted human attention since classical times. Chameleons fascinated Aristotle, and he speculated on how they change color. He also maintained that honey bees build their combs using wax they gather from olive trees. The color of chameleons and the wax of bees are manifestations of lifestyle-chemicals critical to these creatures' lives. Our curiosity is as keen as Aristotle's, and we are now beginning to understand chemical adaptations in detail. On occasion, we can even turn this new knowledge to practical use. The structural materials used by living organisms, for example, offer novel ideas for strong, light-weight construction. Deer antlers, spider silks, and mussel shells are just a few of the materials that scientists have analyzed to learn how nature has solved particular engineering problems.

ANTIFREEZE IN ANTARCTICA

Anyone who keeps a car in a freezing climate knows to put antifreeze in the cooling system. Otherwise, the cooling water will freeze and probably crack the engine block. Antifreeze permits the coolant to remain liquid at lower temperatures and protects against this expensive mishap. (Dissolving antifreeze or any other chemical compound in water causes the water to freeze at a lower temperature than pure water. The greater the amount of chemical dissolved, the lower the freezing point becomes.)

Living creatures face a similar threat in cold weather. Their lives depend on the complex machinery of cells filled with water. If the cellular water freezes, the organisms may die. Warm-blooded animals like ourselves avoid freezing by producing enough body heat to warm themselves above the temperature of their surroundings. With their body heat and various means of conserving it, warm-blooded animals can withstand freezing weather. However, this solution is not available to other organisms. Their internal temperature drops with the temperature of their surroundings. Plants, invertebrates, and cold-blooded vertebrates must find alternative solutions to the problem of freezing. Only a small number of them have evolved ways to continue their lives unchanged in subfreezing cold. How they do this is a fascinating scientific puzzle with substantial practical implications. Understanding how these creatures survive low temperatures may ultimately teach us how to preserve human tissues in the cold. Storage banks of healthy human kidneys, hearts, and other vital organs could revolutionize organ transplant surgery.

Not surprisingly, some of these cold-weather species live on the icy

108

continent of Antarctica. Apart from warm-blooded animals such as penguins, the dominant species are several kinds of mites and some primitive insects known as springtails. These creatures stay alive through winters with temperatures as low as –28°C. There is at least one hardy springtail (*Isotoma klovstadi*) that withstands temperatures down to –60°C.

One mite that functions under extremely cold conditions is *Alaskozetes antarcticus*, a tiny brown-black animal that feeds on algae and detritus. Widely distributed in Antarctica, *A. antarcticus* is common in areas fertilized by birds or seals and in places hospitable to green algae. During the antarctic summer, the mites disperse to hunt for food, but in the winter they often huddle in aggregates of a thousand or more individuals. Each mite is about 1 millimeter long and weighs 200 to 300 micrograms (about as much as an ordinary grain of table salt). Its life cycle is not seasonal, so both immature and adult mites must continue through the Antarctic winter. As the weather grows colder and the temperature drops below 0°C, which is the freezing point of pure water, the mite makes two adjustments. It stops feeding, and it begins synthesizing glycerol, a simple compound chemically very similar to the antifreeze routinely used in automobiles.

The mites synthesize enough glycerol to make a 5-percent solution in their cells. At this concentration, glycerol makes up about 1 percent of the mites' total body weight and is an effective antifreeze. This is sufficient glycerol to prevent ice crystals from forming above about –28°C, but only if the mites do not eat. If the mites have food in their gut, the glycerol is less effective. This is because glycerol affects freezing in two different ways.

Glycerol lowers the freezing point of water, in the same way as automobile antifreeze and other chemical additives do. In addition, it allows water to remain liquid many degrees below the freezing point. It may seem contradictory to speak of a liquid below its freezing point, but many liquids can be carefully cooled below their freezing point without solidifying. Chemists call this unstable condition supercooling.

It is not difficult to supercool water, that is, to keep water liquid below its freezing point, provided the water is free of solid particles. Any bits of solid serve as nuclei around which ice crystals can begin to form and thus prevent supercooling. A small ice crystal introduced into a container of supercooled water serves as an excellent nucleus for crystallization. As soon as the crystal enters the liquid, the liquid quickly freezes solid. (In contrast, a small crystal of ice introduced into water at a temperature just above its freezing point simply melts.)

In the same way, bits of food in an Antarctic mite's gut serve as nuclei around which ice crystals can form, negating the supercooling effect of glycerol. Presumably, this is why the mites cease feeding when the temperature drops below freezing. A mite with an empty gut freezes at a lower temperature than one that has recently fed. It is better to be hungry than frozen.

Supercooled mites and springtails regularly withstand winter temperatures below −20°C on the Antarctic landmass, but the creatures that live in the surrounding oceans never face such extreme conditions. Their temperature never drops below about −1.9°C, the typical freezing temperature of Antarctic seawater. If the weather becomes colder and the air temperature falls, more ice forms in the ocean, but the water temperature remains constant. Fluctuation in the ice mass keeps the water temperature constant at the freezing point as the air temperature changes. (The freezing temperature of seawater is lower than 0°C owing to its dissolved salts. The saltier the seawater, the lower its freezing temperature.) At the relatively warm temperature of −1.9°C, many aquatic microorganisms and invertebrates flourish. Each year, the seas of Antarctica produce about 610 million tons of plankton that serve as food for large communities of other residents. These marine microorganisms and invertebrates survive because the contents of their cells are sufficiently concentrated that they freeze only at a temperature somewhat lower than −1.9°C. They require no antifreeze to thrive.

For vertebrates the situation is different: their cells are much more watery than the cells of microorganisms or invertebrates. Containing a lower concentration of dissolved chemicals, vertebrate cells consequently freeze at a higher temperature than invertebrate cells. The cells of typical marine fishes from around the world freeze at −0.8°C.

The result is that most of the world's fishes die promptly in freezing seawater. A fish taken from temperate Atlantic waters and immersed in the frigid Antarctic Ocean freezes at once. Ice crystals spread visibly through its tissues as their temperature drops below the body's freezing point. Nonetheless, there are plenty of local fish that remain unfrozen in Antarctic waters. When zoologists first visited Antarctica, they were astonished to find fish swimming among ice crystals and even resting on submerged shelves of ice.

We now know that most of these local fishes belong to a single group of about one hundred perchlike species. They survive because, like Antarctic mites and springtails, they protect themselves with an antifreeze. The compounds they use are chemically much more complicated than glycerol. The fishes' antifreeze consists of complex molecules known as glycoproteins, which are proteins bonded chemically to simple sugar units.

Glycoprotein antifreezes prevent water from freezing by blocking the growth of ice crystals, rather than simply lowering the freezing point of the solution. To understand how they function, we should first look at how an ice crystal grows as water freezes. A microscopic ice crystal increases in size by adding water molecules to its outer surfaces from the surrounding liquid water. Each new water molecule fits nicely into a space on the ice surface. The molecule remains in place and becomes part of the growing crystal because its hydrogen and oxygen atoms interact with the atoms of other water molecules already in place on the ice surface. As more water molecules add to the ice surface, the crystal continues to grow at the expense of the surrounding liquid water. Eventually all the molecules of water have found a place on a crystal. At this point, the liquid water has turned totally to ice.

Scientists have not yet unravelled every detail of how glycoprotein antifreezes work, but the overall process is reasonably clear. Small portions of these large molecules fit conveniently into the surface of growing ice crystals, just as water molecules do. These parts of the glycoproteins bind quite snugly to the ice surface and block new water molecules from adding to it. If new molecules of water from the liquid cannot fit into an ice surface, they simply remain as liquid water. Thus, these antifreeze molecules physically inhibit the growth of microscopic ice crystals and prevent cellular water from freezing. Glycoprotein antifreezes protect Antarctic fish down to just below −2°C, a few tenths of a degree colder than the constant temperature of the surrounding sea.

GLYCOPROTEINS are effective antifreezes because they bind to ice crystals. There are other kinds of lifestyle proteins that also work by binding to particular chemicals. All proteins consist of chains of amino acids, and many proteins bind to other compounds through special chemical units attached to the amino acid chain. In glycoprotein antifreeze molecules, for example, it is the sugar units that attach directly to the ice surface. In some cases, the binding groups are part of special amino acids that are incorporated into the protein chain. These special amino acids are different from the twenty common amino acids found universally in proteins.

One lifestyle protein that functions by binding to a particular chemical is important in constructing oyster shells. The common American oyster (*Crassostrea virginica*) constructs its shell largely from calcium carbonate, or chalk. This chemical compound consists of calcium ions and carbonate ions in equal numbers. Oyster shell is significantly stronger than chalk, partly because oysters build their shells with the aid of a protein that binds to calcium ions. This protein determines how and where solid calcium carbonate is laid down as the shell grows, making the structure regular and ordered, and contributing strength to the shell. A related protein in mammals participates in forming dentin, the dense calcium-containing material that constitutes the principal mass of teeth.

NATURAL ADHESIVES

Another group of special binding proteins are the active constituents of biological adhesives. Some sedentary organisms, such as mussels and barnacles, attach themselves to objects in their environment by means of adhesives. Other organisms use adhesives in construction, much as we use mortar to cement bricks or stones together in structures we build. Natural adhesives are chemically unlike the ones that we use, and both biologists and materials scientists are curious about how they work. Much of what we now know comes from studies carried out by J. Herbert Waite and his colleagues at the University of Delaware on the common blue mussel. Mussels anchor themselves to rocks or other underwater surfaces. They are opportunists and not very particular about what they grasp, as long as it gives them a holdfast. On a sandy bottom where firm solid surfaces are rare, a mussel may be satisfied with several small loose rocks.

A mussel attaches itself to its holdfast by means of dark-colored strands known as byssal threads (sometimes called the mussel's "beard"). These threads run between the mussel and the holdfast and look like coarse hairs or fibers. Aristotle was acquainted with byssal threads, but he mistook them for a foreign parasite that had invaded the mussel. In reality, the threads are tendons that originate in the animal's retractor muscles. They terminate in small flattened discs that are tightly bonded to the holdfast by a powerful protein adhesive. If a mussel is pulled from its holdfast, the tough byssal threads usually break before the adhesive fails.

Mussel adhesive owes its strength to proteins that incorporate a special amino acid called dihydroxyphenylalanine, or dopa. This amino acid binds readily with silicon, as well as with many metals, including iron and aluminum. Silicon, iron, and aluminum are extremely abundant elements throughout the earth and are important constituents of terrestrial rocks. As a result, dopa binds strongly to chemical compounds generally found on rock surfaces, and therefore the dopa-containing proteins can adhere to most objects the mussel tries to grasp. Without this unspecialized, broadly effective adhesive, the mussel's lifestyle could not include opportunistic attachment to a holdfast.

Other dopa-containing proteins are key components of the adhesive used by sandcastle worms (*Phragmatopoma californica*) in constructing their castles. These worms belong to a large group of common seashore animals known as marine bristleworms. They are relatives of earthworms and are fairly simply organized, much less complex than mussels. Sandcastle worms are small, gregarious creatures that erect and inhabit extensive reeflike mounds or "sandcastles" in shallow intertidal waters. Wave action is unavoidable in the intertidal zone, and for the mounds to remain habitable, they must withstand the persistent pounding of the tides. The mounds are honeycombs of contiguous tubes, open at one end and stacked upon one another like rooms in an apartment house or bottles on their sides in a case. Each worm builds itself a tube from small grains of sand and other particulate matter that it gathers from the water, cementing these building blocks together by means of an adhesive that it secretes.

As the worm grows, it enlarges its quarters by cementing additional bits of sand into place. It remains within its tube and feeds by extending long slender tentacles into the open water to sieve for plankton. Whenever a predator threatens, the worm hastily retracts its tentacles, retreats into its tube, and plugs the open end with a cover attached to its head. It also seals its tube in this way to avoid desiccation if low tide leaves the mound high and dry.

The worms' distinctive way of life depends on their mounds, and in constructing the mounds they obviously depend upon their adhesive. The way they handle this adhesive is quite remarkable, especially when compared with our own cumbersome methodology for applying adhesives. In our procedure, we first prepare the surfaces to be joined by cleaning them and then covering them with a coupling agent. Next, we spread a layer of adhesive. Our adhesives are chemically complex formulations of a monomer, an initiator, a cross-linker, and perhaps a catalyst and other ingredients. We finally bring the surfaces together and immobilize them while the adhesive sets. While we might sometimes find it useful to cement one solid to another underwater, our methods make this task inconvenient, if not impossible.

In contrast, sandcastle worms build sturdy houses with no concern about cleaning or preparing the particles they join, and they routinely use their adhesive underwater. It bonds tenaciously to a variety of materials; it is effective in relatively small amounts, and the worms produce it as needed. Many materials scientists would be pleased to patent such a glue. These simple worms have an adhesive that we cannot equal and do not yet fully understand.

If we are to discover the secrets of marine adhesives, we must analyze the chemical events that take place when an adhesive joins two surfaces. To do this, we must first obtain the adhesive proteins responsible for bonding and learn their chemical structures. A significant step in that direction came in 1992 when Waite, working with Rebecca A. Jensen and Daniel E. Morse of the University of California at Santa Barbara, isolated two dopa-containing proteins from the sandcastle-worm adhesive.

Waite, Jensen, and Morse obtained these adhesive proteins from about two hundred sandcastle worms they collected along the Pacific coast near Santa Barbara. They removed the thorax, which contains the adhesive, from each worm, and ground the two hundred thoraces to a fine paste. They extracted the two proteins from the ground tissue and purified them following standard biochemical procedures. Their task now is to unravel the complete chemical structures of these compounds and begin learning how the adhesive functions.

Biological adhesives are an intriguing puzzle, but interest in them goes beyond scientific curiosity. There is a real possibility of exploiting natural adhesives in designing better adhesives for human use. Once materials scientists understand how biological adhesives work, they should be able to create synthetic adhesives that mimic the natural counterparts. The possibility of improved surgical, dental, and orthopedic adhesives is particularly attractive. Potential industrial applications include simpler methods for joining surfaces and anticorrosive compounds that bind firmly to metal surfaces. We might even get an underwater glue!

SILKWORMS AND SPIDER SILK

Unlike biological antifreezes and natural adhesives, most proteins do not contain sugars or special amino acids to bind other chemicals. One group of proteins made up of only ordinary amino acids forms a very important family of lifestyle chemicals. These proteins have their amino acids arranged in a unique way that creates strong, thin fibers. (The properties of proteins, like other molecules, result from their specific molecular structures. The characteristics of a particular protein arise in a complicated way from which amino acids are joined, and in what order, to make the protein.) All the proteins of this particular family of lifestyle proteins form fibers, but their specific properties vary considerably, so that these fibrous proteins serve a variety of purposes. In practical human significance, the leading member of the family is the protein that is the source of textile silk, and collectively these fibrous proteins are called silks.

Textile silk is doubtless the commercially most important product derived from any lifestyle chemical. Silk is an ancient fabric, with a complex history spanning about five thousand years. Throughout these millennia, silk has represented beauty and luxury, and has always been fashionable among those who could command the best. Half a century ago, nylon replaced silk in parachutes and women's stockings, and other silklike synthetics have since become widely available. None of these fibers has ever matched the feel, or the stylish appeal, of silk.

The fibrous protein responsible for this elegant material is one of the proteins with which the larva of the domesticated silkworm moth constructs its cocoon. (The silkworm moth is nearly always *Bombyx mori*, although a few other species are cultivated in some parts of the world.) Like other moths and butterflies, the silkworm moth undergoes complete metamorphosis. After hatching from its egg, the larva typically feeds on tender leaves of the mulberry tree (*Morus alba*, usually) for about thirty-three days. It is then a fully grown caterpillar and ready to undergo the next step in its development. It begins to construct its cocoon. For the next two days the silkworm spins, moving its head in a figure-eight and emitting two fine filaments from an opening near its mouth called a spinneret. Then it turns head to tail and spins for two days more. Owing to the figure-eight motion, the white mass formed as the filaments dry takes the shape of a cocoon surrounding the silkworm.

When the cocoon is complete, the silkworm is wholly encased. Protected from mold and damp, it then begins its metamorphosis. In this recurring miracle, it turns from a caterpillar into a quiescent pupa that reorganizes itself into an adult moth. Adults emerge from their cocoons to mate, lay eggs, and die. Many groups of insects undergo complete metamorphosis, but only a few enclose their pupae in heavy silken cocoons.

The filaments extruded by the spinning silkworm consist of two proteins. One is called fibroin, and it forms the body of the filament. The other, called silk gum or sericin, coats the two filaments coming from the silkworm and causes them to stick together. The filaments coalesce into what appears to be one fiber. Although generating this fiber is called spinning, the process involves no twisting. The entire cocoon consists of a single fiber, 600 to 900 meters (2,000 to 3,000 feet) in length. Owing to its coating of silk gum, the fiber sticks to itself and gives shape and rigidity to the cocoon.

For a silkworm, this cocoon is a protective case. For silkmakers, this same cocoon is the immediate source of textile silk. In one procedure for making silk, a silkmaker first heats cocoons in a mildly alkaline solution that softens them and removes the silk gum. A worker then grasps the end of the fibroin filament, combines it with fibers from several other cocoons to form a thread, and reels the thread from the cocoons. The silkmaker twists threads together into strands of raw silk.

Some version of this simple process has been in use for the thousands of years that the silkworm moth has lived as a domesticated species. Selective breeding has created more than one thousand commercial varieties of silkworm and has enhanced the quality of its silk. Like other domesticated species, the silkworm moth long ago traded an independent but uncertain existence for the security of a more circumscribed way of life. The species has now lost the ability to live on its own and can survive only in association with humans. Adult male moths can no longer fly, although their wings flutter excitedly in the presence of a female. The larvae consume prodigious quantities of mulberry leaves in their growth phase, but they are incapable of locating their food. Unless they are placed on the leaves, they simply die. Even then, the youngest larvae can feed only if the leaves are first finely shredded.

Domestication of the silkworm moth brought changes in the lives of its discoverers as well. Silk and the silkworm have influenced human affairs for millennia since their discovery in ancient China. Historical details of the origin of silk are lost, but Chinese legend credits Lady Hsi-ling, principal wife of the legendary emperor Huang-ti, with discovering silk around 2600 B.C. Physical evidence supports this date surprisingly well. The earliest-known silk textiles are small fragments of cloth from an archeological excavation in the Chinese province of Chekiang, just south of Shanghai. Radiocarbon dating indicates that these bits of silk originated around 2750 B.C. The silk appears to have come from the domesticated silkworm, and the fragments represent more than one style of weaving. We know then that by about 2750 B.C., the Chinese were making silk. Perhaps raising silkworms was well established by that time.

Whenever it began, ancient Chinese silkmaking was a sacred and awe-inspiring activity. To raise silkworms and harvest their cocoons, the silkmakers had to be thoroughly knowledgeable about the moth's life cycle. They harvested most cocoons for silk but allowed a few insects to complete their metamorphosis so that some adult females could mate and lay eggs for the next generation of silkworms. Although they watched the miraculous transformations again and again, the insect's life cycle remained an incomprehensible mystery in this ancient world. To the discoverers of silk, only a divine miracle could explain these events and the fiber they yielded. Believing that silk was a gift sent by the gods, the Chinese surrounded it with myth and ritual. It is no surprise that insect metamorphosis elicited awe and wonder in these late Neolithic people; it still seems amazing today.

Chinese legends explaining silk glorified Lady Hsi-ling as the first cultivator of silkworms. The Silkworm Goddess occupied an honored place in the Chinese pantheon, and sacrifices at her shrines assured the silkmakers' success. These sacrifices must have been solemn rites, because inscriptions from the twelfth century B.C. record offerings of human captives to her. In these early years, the silkworm cocoon also acquired religious significance. It was a physical reminder of the endless cycles of the insect's life and so came to symbolize resurrection. Jade amulets in the form of silkworm co-coons survive from the second millennium B.C.

For centuries, the ruling class in China reserved silk for its own use, but with the passage of time silk assumed commercial importance and spread beyond Chinese borders. By the ninth century B.C., silk was widely enough available to be used as a medium of exchange. By 600 B.C., Chinese silk reached Europe, where it was treasured for centuries as an exotic luxury. Not until a thousand years later did silkworms themselves reach the Mediterranean.

ALTHOUGH silkworms spin the silk we value most, many other insects also make silk. The most versatile silk-spinners of all, however, are not insects but spiders, whose silks are also fibrous proteins. Both insects and spiders are arthropods, but they belong to different classes of this very large phylum of animals and are only distantly related. Insects and spiders look somewhat similar but are easily distinguished. Spiders have eight legs and no antennae, while insects have six legs and do have antennae. Spiders have no wings, but most insects have one or two pairs of wings. The structure of their bodies and eyes is also different. Spiders belong to a class of arthropods called arachnids, which also includes the ticks, mites, and scorpions. Biologists who have reconstructed their history estimate that the earliest spiders lived about 380 million years ago.

All spiders spin silk, and they generally begin to do so shortly after hatching. In addition to the familiar web-spinning spiders, there are spiders that follow quite different lifestyles, but all of these depend in one way or another on silk. Some spiders live underground in silk-lined burrows. Others (*Mastophora* species) capture flying insects by hitting them with a blob of sticky silk that they swing on a silken tether. (The discoverer of one of these silk-throwing spiders was so impressed with its aim that he named it *Mastophora dizzydeani.*) Some small spiders live in the corners of larger spiders' webs and steal food from their hosts.

Many spiders spin several kinds of silk for various purposes throughout their lives. Spiders wrap their eggs in silk and employ a silk cement. Some spiders build webs requiring more than one kind of silk in their construction. Spiders also travel on silk, lowering themselves on silken strands called draglines and passing back and forth on silken bridges.

Spiders even sail through the air on silk. This is known as "ballooning," and it is widespread in many families of spiders. Young spiders float far from home, scattering themselves through the countryside by the thousands. On a fine day when there is no more than a gentle breeze, a spiderling climbs to a high point. It stands head down on all eight legs, pointing its abdomen into the air. Its silk glands and spinnerets are in the rear of its abdomen and so are now aimed upward. In this position the spiderling begins to release silk from its spinnerets. Spiders do not discharge silk with force, but the breeze carries the spiderling's fresh silk away as it appears. Soon there is enough for a gossamer airship, still attached to the spider but fluttering in the breeze. The small spider lifts its legs from its headstand, reaches up to grasp the waving strands, and sails away. The journey is perforce downwind, but the spiderling has some control over the distance it travels. If it wishes to land, it merely climbs up the strands of silk and begins rolling them into a ball, causing the airship and its passenger to descend to earth. Not every

6.1 A spider prepares to travel on the wind. Most ballooning spiders go no higher than about 70 meters (nearly 230 feet), but they occasionally reach altitudes as great as 5 kilometers (over 16,000 feet).

spiderling ends its flight this way, and some take quite long journeys. Ships at sea occasionally meet ballooning spiders hundreds of miles from the nearest land.

Spiders can spin a variety of silks because they have several distinct silk glands. Each gland synthesizes a chemically different silk protein and yields its own characteristic silk. The spider switches from one gland to another, depending on the silk it requires. The female of the everyday British garden cross spider (*Araneus diadematus*) calls upon seven different silk glands to satisfy her needs.

In addition to synthesizing chemically different silks, some spiders can modify the diameter of the fiber they spin and also vary its physical properties, such as strength and elasticity. Biologists believe that spiders do this by manipulating the valves and machinery of their silk-producing system. Silk production starts with silk proteins, which the spider makes and stores in solution in its silk glands. When the spider spins silk, these proteins pass down a small tube from a silk gland to a spinneret and then out through a spigot. Passage down this tube is critical, because during this journey the protein molecules become aligned in a way that converts the soluble proteins into a strong, insoluble fiber. How this alignment comes about is not completely clear, but it requires a change in shape of the protein molecules. The spider apparently controls the process to produce silk with variable physical properties.

Control over their silk permits spiders to spin fibers remarkably well adapted to various needs. A fine example is the familiar spider web consisting of a set of radial spokes supporting a flat, closely spaced spiral, one of nature's most beautiful constructions. Biologists call this an orb-web, and like other webs, its primary purpose is to trap insects for food. An efficient orb-web must stop a flying insect and retain it until the resident spider arrives to inspect the new captive. If the web were brittle and weak, the insect would fly straight through. Like a bullet shot through a sheet of paper, it would leave only a hole to mark its passage. If the web were elastic but recoiled rapidly after being stretched, it would behave as a trampoline. An insect striking it in mid-flight would simply bounce off.

An orb-web is an engineering marvel that has neither of these shortcomings. The spider spins firm, dry dragline silk for the spoke threads and web supports. After laying down this strong radial framework, it switches to a sticky, softer silk and spins a single thread in a long spiral over the spokes. The spider starts this spiral at the periphery of the web and works toward the center, frequently reversing direction as it spins. This spiral is the capture-thread. It adheres to the stiffer spoke threads, but its softer silk can stretch considerably without weakening.

When it is complete, the orb-web may weigh less than half a milligram, the weight of a couple of grains of salt, and yet it can comfortably support a spider weighing a thousand times as much. It absorbs a flying insect's energy without flinging the insect back into the air. It is elastic enough that an insect does not pass through it but becomes entrapped in its strands. A trapped insect struggles to free itself but cannot easily gain a purchase on the soft, tacky capture-thread. Often the captive fails to extricate itself before the spider arrives to inspect its latest acquisition.

Spider silk is suited to the spider's needs and is also exceptional in comparison with other materials. On a weight basis, more energy is required to break dragline silk than high-tensile steel. Dragline silk is also stiffer than other familiar biological building materials, such as cellulose fiber, tendon, or bone. These properties make spider silk an attractive material for human use, and it has received serious scientific attention. For many years scientists made little progress in understanding the characteristics of silk, but several research groups are now changing this situation.

One group of biologists working on spider silk is a team led by Randolph V. Lewis at the University of Wyoming. These biologists are studying golden-silk spiders (*Naphila clavipes*) and have made significant advances on several fronts. They have learned to draw dragline silk from an anesthetized spider and have discovered that this silk consists of two different proteins. They have worked out the chemical structures of these two proteins and are now exploring the chemical basis of this silk's unique properties. The United States Navy has supported this research under a program devoted to basic materials science. In related research, the United States Army is studying dragline silk as part of a search for strong, lightweight materials for bulletproof vests.

Fundamental studies such as these relate spider silk's properties to its physical and chemical structure. With this information, scientists should be able to create new synthetic fibers that have the desirable properties of natural spider silks. In addition to this approach, materials scientists are pursuing another quite different idea here. They are examining the feasibility of producing spider silk itself on a large scale.

The idea of producing spider silk commercially has been around for at least three hundred years, but a successful venture has yet to emerge. Among other problems, no one has devised a realistic scheme for raising and feeding the large numbers of spiders required. Because of the spiders' way of life, this would be a prohibitively expensive undertaking. Unlike silkworms, spiders are belligerent carnivores. Instead of the mulberrry leaves that silkworms crave, spiders demand fresh insects to eat. Silkmakers can crowd many silkworms into small cages, but most species of spiders kill one another if they are housed together.

It now seems possible that the practical obstacles to spider-culture can be circumvented. Future research may open the way to producing spider silk without raising spiders at all. This is less implausible than it sounds, because modern molecular biology permits scientists to manipulate the genetic material (DNA) of organisms. Using genetic engineering techniques, a molecular biologist can isolate a gene, which is a piece of DNA that carries chemical instructions for a cell to follow, from one organism and make it part of the DNA of another species. In many cases, the transferred gene functions normally in the second species. Here is how it could work to produce spider silk.

Lewis's group has already employed molecular-biological techniques to identify the genes for the two dragline proteins from golden-silk spiders. From spider cells these biologists isolated two pieces of spider DNA that carry the instructions for synthesizing the dragline proteins inside a cell. They introduced these spider genes into the DNA of laboratory bacteria, so that these bacteria now carry the spider genes along with their own genetic material and pass the spider genes on when they reproduce. In principle, the modified bacteria should be able to read the instructions provided by the spider genes and synthesize the dragline proteins. Lewis's initial experiments here were only partially sucessful. The modified bacteria did not read the genes' instructions very well and made only tiny amounts of the spider proteins. With future improvements, the biologists should obtain bacteria that synthesize the dragline proteins efficiently.

If these experiments are successful, these bacteria could become a convenient source of spider silk. Producing spider silk commercially then would be a matter of cultivating bacteria in large vats. Five thousand years ago, silk from silkworms was a miracle; in the future, spider silk from bacteria may be a reality.

THE FIRE OF FIREFLIES AND GLOW OF GLOWWORMS

Fireflies have delighted people the world over since time immemorial. They appear in the Upanishads, written by Indian sages before 600 B.C., and even earlier a Chinese writer mentions them in a manuscript from the second millennium B.C. Few other familiar creatures generate their own light, and the little flashes that fireflies make still fascinate us today. Light generated by fireflies is called bioluminescence. Though the best known to us, fireflies are only one of several kinds of luminescent creatures that have attracted attention since classical antiquity. Aristotle knew about luminescent fishes, mushrooms, and glowworms in addition to fireflies, and the Romans recognized even more examples.

Bioluminescence has also intrigued scientists since the earliest days of careful experimentation. Robert Boyle, the great seventeenth-century English investigator, wrote of his "wonder and delight" on seeing luminescent creatures. Boyle is probably best remembered today for his work on the behavior of gases, but the study of glowworms that he described in 1672 was one of the earliest investigations of a biochemical reaction. Boyle found that a glowworm's light depends on the presence of air. He placed two brightly shining glowworms in a sealed chamber. When he pumped air out of the chamber, the glowworms soon ceased to glow. When he let air back in, their light returned. Excised light organs behaved similarly.

Boyle did not attempt to explain his observations; in fact, three centuries would pass before anyone could. Glowworm luminescence depends on air because it is a chemical process that consumes oxygen. Unlike the radiant filament of an ordinary light bulb, the light of luminescent organisms is the product of chemical reactions. An incandescent light bulb gives off light because the filament is very hot, over 2000°C. Fireflies and other luminescent creatures could never produce light in this way. The high temperature necessary for incandescence would quickly burn up their bodies. A firefly that used heat to make light would self-destruct on its first flash.

Luminescent creatures use chemical reactions that generate light at ordinary temperatures. When chemicals react, new molecules are formed.

These new molecules often have excess energy that they must eliminate quickly. (We encountered the need to get rid of excess energy earlier, when we talked about hypericin. The hypericin molecules acquired their excess energy by absorbing sunlight and then passed it on to other molecules.) Newly formed molecules generally dispose of their excess energy as heat. When energy appears as heat, the local temperature rises. This is what happens to most of the energy released when gasoline reacts with oxygen in an automobile engine. However, a few kinds of chemical reactions are different. They also generate molecules having excess energy, but the molecules dispose of this energy as light. Instead of raising the local temperature, these reactions release a flash or shimmer of light. Luminescent organisms have learned how to carry out light-producing chemical reactions in their cells. There are also many organisms, particularly fishes, whose light is not really their own but is generated by colonies of luminescent microorganisms they support.

The various reactions responsible for bioluminescence require a specific compound, called a luciferin, and a specific enzyme, called a luciferase. Chemists have obtained several chemically unrelated luciferins and luciferases from luminescent organisms. Although these components vary, most light-producing biochemical reactions follow the same basic course. The luciferase catalyzes a chemical reaction between the luciferin and oxygen molecules present in the organisms' cells. This reaction leads to a new compound that is unstable and reacts further to produce light. Overall, there may be a sequence of several chemical transformations, but their net effect is to consume luciferin and oxygen, and to generate light. This fundamental scheme applies to fireflies, glowworms, and other luminescent species that chemists have examined in detail. It also clarifies Robert Boyle's observation that glowworms glow only when air is present. Glowworms need oxygen in their cells to luminesce, and cellular oxygen comes from the atmosphere.

Organisms all across the biological spectrum have mastered reactions that give off light. Some shine continuously; others regulate the luciferin reaction and turn their light on and off. Bioluminescence is widespread in some animal phyla, but about half the phyla have no known luminescent members. For example, bioluminescence is prevalent among marine fishes but totally absent in all higher vertebrates. It is common in beetles, which include the firefly family, but rare or absent in several other groups of insects. We know of some luminescent earthworms and shrimps, many luminescent squids and jellyfishes, but no luminescent scorpions or salamanders. On the other hand, luminescent microorganisms are widespread, particularly in the oceans. Marine bacteria and dinoflagellates frequently illuminate the seas at night. Sometimes they appear massed offshore near the surface, where swimmers can pass through their luminous clouds and emerge radiant and shedding sparkling drops of water.

Luminescent organisms seem to be scattered among the species in an uneven distribution without rhyme or reason. This apparently random distribution is one of the reasons many biologists believe bioluminescence has arisen independently a number of times as life has developed on earth. Other evidence on the origin of bioluminescence is confusing, however, and questions about how and when organisms learned to make light remain unresolved.

What biologists understand better are the advantages of generating light. Some organisms use light to communicate with other members of their species. Male fireflies, for example, seek their mates with coded flashes. Others creatures, such as luminescent dinoflagellates, use light to defend themselves against predators. Although luminescent marine microorganisms have been familiar at least since classical times, it was long uncertain what the advantage of luminescence was for these creatures. Modern biologists settled this question with a clever laboratory experiment. They mixed two related species of dinoflagellates, one luminescent and the other not, in a tank and added predators that feed on both species. In this mixture, the luminescent dinoflagellates fared better than their dark cousins. Apparently, the luminescent species flashes when attacked, momentarily distracting or frightening the predator and increasing its own chances of escape. As a result, it survives longer than the nonluminescent species.

Various other organisms also use luminescence for defense, but sometimes the aggressor, rather than its prey, has the light. Some luminescent predators use light to locate their prey. Others, such as glowworms, use light to lure unwary prey to themselves. Glowworms are the carnivorous larvae of certain insects, including small flies known as fungus gnats that live in damp environments around the world. Biologists have described about two thousand species of fungus gnats, and about a dozen are luminescent. The most exhaustively studied of these is *Arachnocampa luminosa*, a New

Zealand gnat that lives along shaded streams, in rainforests, and in tunnels and caves.

One of this gnat's best-known habitats is a cave near Waitomo on New Zealand's North Island that tourists have been visiting for decades. Gnat larvae live by the thousands along the ceiling and walls of this cave, each about 1 to 1.5 centimeters (about 0.5 inch) long and furnished with a tiny knob of light at the end of its body. The larvae extinguish their lights when frightened, but otherwise an eerie greenish glow penetrates the subterranean darkness.

Shortly after a larva hatches, it spins a tubule of silk and suspends a number of long silken threads from it. The tubule rests against the ceiling or wall of the cave and serves as the larva's home. The threads are up to 50 centimeters (20 inches) long and as strong as spider silk. Each one is dotted with clear droplets of gluelike mucus. The mass of threads forms an efficient snare that dangles below the tubule. The larvae's lights attract midges and other insects that breed on the floor of the cave. They fly toward the light and become entangled in the sticky hanging threads. As soon as a larva senses an insect struggling on one of its lines, it emerges from its tubule and hauls in the line to retrieve its prey. When they are not feeding on midges, the larvae often attack one another. They are pugnacious, cannibalistic little creatures, and two larvae fight to the death with their lights brightly shining. The winner devours the loser but discards its light. The little knob continues to glow as it falls to the floor of the cave.

ATTRACTING prey with light is unexceptional among luminescent organisms. A less obvious way to use light is prevalent among many luminescent fishes that inhabit the middle levels of the ocean. These creatures have turned their light into a subtle form of camouflage. Like other organisms, they benefit from making themselves inconspicuous to predators. Color often reduces their visibility from above, but a fish in this dimly lit environment has another problem that color cannot solve: when viewed from below, a mid-level fish appears as a dark silhouette against the faint light filtering down from the water's surface. No matter what its markings, it remains visible and exposed to attack from below.

A fish's dark silhouette invites hungry predators, but there is a way to minimize this risk. A fish could make its silhouette disappear by replacing the light that its body blocks out from above. If a fish's under, or ventral, surface glowed evenly with the brightness of the natural light from above, the fish would have no silhouette. A glowing fish would blend in with its surroundings and become less vulnerable. Many luminescent fishes thus have evenly illuminated ventral surfaces that provide camouflage of this sort.

For the deception to be most effective, a fish's light should exactly match the natural light from the ocean surface, both in angular distribution and in brightness. Under these conditions, the fish's luminescence appears indistinguishable from the surrounding natural light. To probe the angular distribution of light, scientists at the laboratory of the Marine Biological Association of the United Kingdom under the leadership of E. J. Denton made careful measurements on two species of luminescent marine fishes, hatchet fish (*Argyropelecus affinis*) and viper fish (*Chauliodus sloani*). They found that the light from the ventral surface of these fishes has the same angular distribution as the daylight from above. That is, the intensity of light in various directions coming from the surface of the fish matches that of the surrounding light coming from above. Other biologists have observed that several species of luminescent fishes monitor the local level of natural light from above and adjust their own brightness to match the surroundings. The ability to adjust the light level is important to a fish, because natural light in the ocean becomes dimmer at greater depths and changes with the weather, season, and time of day.

These mid-level fish have developed a single critical application of luminescence. A fish that takes a different approach to using light is the flashlight fish. Like a person who carries a flashlight wherever he goes, this fish has a light that is always available for a variety of purposes. It is one of many fishes that do not generate their own light but rather depend on colonies of luminescent bacteria that flourish in the fishes' light organs. The fishes furnish the bacteria with food and a stable environment, and the bacteria obligingly serve as the fishes' flashlights.

Four closely related species are called flashlight fish, but one of these (*Photoblepharon palpebratus*) has received most of the scientific attention. This is a dark, reclusive animal, about 8 centimeters (3.5 inches) long, that marine biologists have encountered from the Red Sea eastward to Indonesian and Japanese waters. Flashlight fish may be even more widely distributed, but their retiring habits make them difficult to find. They are most active on moonless nights and spend the daylight hours in deep caves and recesses, hidden from view. In most habitats, they remain at a depth of about 30 meters (100 feet).

A flashlight fish has two light organs, each about 1 centimeter long and 0.5 centimeter high (0.4 by 0.2 inch), placed horizontally, one beneath each eye. These organs are chambers densely packed with luminescent bacteria. The bacteria glow continuously, but the fish limits the illumination emitted by means of eyelid-like shutters that it raises to cover its light organs and extinguish the light. Each fish produces a greenish light comparable to that of a weak flashlight. These curious creatures have attracted scientific attention for more than two hundred years, but most of our information on how they exploit their luminescence comes from modern studies carried out independently by James G. Morin of the University of California at Los Angeles and John E. McCosker at the California Academy of Sciences in San Francisco. Their research required laboratory experiments, extensive observations in the aquarium, and nighttime dives in the ocean—efforts that have revealed a lifestyle uniquely dependent on light.

Among luminescent creatures that can regulate their light, most flash (turn their light on) only occasionally. In contrast, flashlight fish keep their shutters open most of the time but occasionally blink (turn their light off). A flashlight fish blinks more often when it sees other flashlight fish, yet it does not respond to the presence of other species. This difference suggests that flashlight fish communicate with one another by means of light signals. Their light assists them in predation, as well. They regularly feed on planktonic organisms that are attracted to light, so the illumination from a group of flashlight fish draws their prey. The light also enables them to see prey. In the laboratory, lightless flashlight fish fail to find prey added to their tank in the dark, although normal specimens have no difficulty.

In addition, light helps flashlight fish avoid their own predators. The luminescence probably lights up the predators, making them more visible and therefore less dangerous. An aggregation of brightly shining flashlight fish may intimidate a predator more than a group of dark fish would. Flashlight fish also employ a "blink-and-run" swimming technique that must confuse pursuing predators. When a fish swims through an open area that offers no protection, or when it is disturbed by divers or predators, it swims with its light on, then blinks the light and at the same time abruptly changes direction. When the light reappears, the fish is in an unexpected location. This blink-and-run swimming pattern is continuous, and presumably the evasive maneuver makes the fish difficult to track.

The flashlight fish uses its light for communication, predation, and defense. As far as we know, luminescent organisms have not discovered other ways to use their light, and most of them use it for only a single purpose. Flashlight fish are the only creatures we have discovered that have developed all three of these functions of bioluminescence. They have an entire lifestyle built around their blinking lights.

$\mathcal{F}$OR OVER three hundred years, biologists and chemists have pursued inquiries into bioluminescence out of curiosity, sharing the sense of "wonder and delight" that Robert Boyle first expressed. For much of this time, practical application of their findings did not appear likely, both because obtaining functional luciferins and luciferases from luminescent organisms is expensive, and because these chemical compounds are unstable and sensitive. In recent years this picture has changed. While bioluminescence has not moved into the public marketplace, it has contributed a convenient technique to the biochemical research laboratory. Biochemists have developed ways of using the firefly luminescence reactions to analyze for particular chemicals. That is, they make use of the luminescence reactions to measure how much of a particular chemical is present in a sample of interest. Because it is relatively easy to measure small amounts of light accurately, bioluminescence has become a preferred technique for the ultrasensitive analysis of small amounts of several important biochemicals. Firefly luciferin and luciferase are now commercially available laboratory chemicals.

Molecular biologists have investigated a second laboratory application of bioluminescence. In transferring genes from an organism into bacteria, only a small percentage of the bacteria may incorporate the genes into their DNA as desired. One of the critical steps in the procedure is identifying the bacteria that have successfully incorporated the genes after an attempted transfer. The investigator must test the bacteria to find out whether the new gene is present.

Biologists at the Agouron Institute in La Jolla, California, have used bioluminescence to solve this problem. They prepared a piece of hybrid

DNA that combined the gene of interest to them, called the target gene, with genes controlling bioluminescence that they took from a luminescent marine bacterium. They transferred this hybrid DNA into laboratory bacteria. Because all the transferred genes were on a single segment of DNA, bacteria that successfully incorporated the target gene also incorporated the luminescence genes. The luminescence genes caused the laboratory bacteria to luminesce, so that the biologists could identify bacteria now carrying the target gene simply by determining which bacteria gave off light. There are other ways to test for successful incorporation of a target gene, but this method proved to be particularly fast and convenient.

COLOR FOR CONCEALMENT AND FOR PUBLICITY

The bright colors of the living world are all around us. Perhaps flowers come first to mind, but the colors of untold numbers of other living things surround us as well: butterflies, mushrooms, ladybugs, hummingbirds, and thousands more. There are also hidden colors that we rarely see. In distant oceans, fishes in brilliant hues dart among delicately tinted coral reefs; gorgeous nudibranchs graze on pastel sponges and flowerlike sea anemones.

The glorious colors of the living world are due largely to chemical compounds called pigments. Living organisms either make pigments for themselves or appropriate them from others, much as they acquire chemical warfare agents. Because color critically influences an organism's way of life, pigments have received the same sort of evolutionary trial and refinement as other lifestyle chemicals over eons of biological development.

Natural pigments fall into about ten chemical families. The members of each family share the same basic chemical structure, but each member differs from the others in its family by small structural variations. Different structures lead in turn to gradations of color among the pigments in a family. One family of flower pigments ranges from pink to red to purple. Another set widely distributed in plants, birds, mollusks, and elsewhere covers the range of yellow, orange, and red.

Gradations of color within a pigment family arise because a compound's color is a direct consequence of its specific chemical structure. This connection between color and structure comes about in the following way. As we mentioned in discussing hypericin, chemical compounds absorb light. "Light" includes both the portion of the spectrum that is visible to humans (what we call "visible light"), and also portions of the spectrum that humans cannot see, such as ultraviolet light. In talking about hypericin, we focussed on what happens to the light energy absorbed by a compound. When pigments absorb light, this energy is usually dissipated as a very small amount of heat. In considering pigments, we are particularly interested in the wavelengths of light that a compound absorbs. Every chemical compound absorbs only specific wavelengths of light, and these wavelengths are determined primarily by the compound's structure. If a compound absorbs wavelengths of visible light, then it appears colored to us. Compounds that absorb different wavelengths of visible light are different colors. Put another way, a pigment appears a certain color to us because it absorbs certain wavelengths of visible light. The slight differences in chemical structure among members of a family of pigments lead to differences in the wavelengths of visible light each pigment absorbs. In this way, a family of structurally related pigments covers a range of related colors.

Whereas pigments absorb visible light, many other chemical compounds, such as aspirin and table sugar, absorb only light that we cannot see. Because they absorb no visible light, they appear colorless to us. However, not all creatures see exactly the same range of wavelengths. Visible light for a bee is not quite the same as visible light for a human. For this reason, chemical compounds that are colorless to one species may appear colored to another.

Coloration supports lifestyles in a number of ways. The role of an organism's color may be apparent only when the organism is in its natural setting. In the zoo, a tiger's stripes make it a gorgeous, spectacular beast. In its natural setting, these same markings help hide the tiger from view. The color we find so striking provides important camouflage to the tiger. Many other organisms, both prey and predator, also want to conceal themselves from the world. The less others notice them, the better off they are. Many ground-dwelling insects, snakes, and birds go about their business arrayed in combinations of brown, gray, and black that make them inconspicuous. Those that live in grass or on foliage may clothe themselves in appropriate shades of green.

Some creatures alter their camouflage as they mature. The fawns of deer as geographically separated as American elk (*Cervus elaphus*) and Chinese water deer (*Hydropotes inermis*) start life with spotted coats that they shed after a few months. The spots make the young fawns less visible when they are resting in grass or in the shade of trees. As they grow and modify their habits, this camouflage becomes less relevant, and it is lost. Similarly, the young of some birds have disruptive down-patterns that later disappear. This protective coloring is common among young birds that leave the nest to run and fend for themselves soon after they hatch. Other young birds are helpless and nest-bound, and typically they do not develop protective coloring.

An organism whose individual development includes a change in life-style often changes its color appropriately. When very young, the larvae of the African moth *Triloqua obliquissima* band together to feed on the surface of a leaf. The small larvae are chalky white, and grouped together on the leaf they look very much like bird droppings. As the larvae grow older, they separate. They now feed individually at the bases and edges of leaves, and each larva rests alone on twigs and leafstalks. To accompany this newfound independence, the larvae change their appearance. They turn from chalky white to a rich shade of brown with greenish splotches, a scheme that nicely matches their new surroundings.

In some species, protective coloration follows the seasons. Many boreal birds and mammals wear earth colors from spring through autumn but turn white with the approach of winter. Ptarmigans (*Lagopus* species), snowshoe hares (*Lepus americanus*), stoats or ermine (*Mustela erminea*), and many of their neighbors undergo this transformation that renders them less visible in a snowy landscape.

Other species do not follow the seasons but adjust their color as they move from one location to another. To gain their living, crab spiders (*Misumena vatia*) lurk among the petals of yellow and white flowers and ambush insects that come to feed. Biologists have found that these spiders alter their color to match their surroundings. A yellow specimen faded to white within a week when it was transferred to a white flower. White spiders placed on yellow flowers gradually made the reverse transition over a period of ten to twenty days.

Crab spiders' shifts between yellow and white are slow in comparison with the abilities of certain reef fishes. Some of these, like the ordinary Nassau grouper (*Epinephelus striatus*) are true quick-change artists. A grouper can assume six or eight different hues and switch from one to another in a few moments, from dark brown to pale gray to a barred pattern to pink. Some shifts are associated with fighting or feeding, but others serve to improve the grouper's camouflage as it moves about.

CONCEALMENT is a sensible strategy for avoiding enemies, but a large number of living organisms prefer bright coloration that attracts the world's attention. They want to be seen, and they advertise their presence to anyone that is looking. The best known of these shameless exhibitionists are flowering plants. As everyone knows, flowers offer food to visitors that they attract with their colors. (Scents are also important here, and we shall return to them later.) These flowering plants cannot reproduce without assistance, and they advertise for the help they need. Birds and insects come to collect the offered nectar and grains of pollen, and inadvertently pollinate the flowers as they feed. After pollination, these plants can set seed and reproduce.

Honey bees are among the most faithful visitors to flowers, but they do not treat all flowers alike. They have definite preferences among flower colors. Bees like blue and yellow but are insensitive to red. That is, they like what we see as blue and yellow. We have no way of knowing how these colors appear to bees. Hummingbirds, on the other hand, distinguish only red among the colors visible to humans.

The colors that flowers offer are under the control of only a few groups of pigments. Many flowers whose colors are in the range from pink, through orange, red, mauve, and purple, to blue owe their color to a single family of pigments known as anthocyanins. The precise color of each anthocyanin depends on whether it binds to other, colorless compounds that may also be present in the flower. This chemical fine-tuning, along with varying acidity in the flower cells, yields a range of shades for each anthocyanin pigment. In this way, although there are only about six common anthocyanins, these pigments provide a wide spectrum of flower colors with subtle variations in shade and tone.

Color attracts an insect to a flower. Once the insect has landed on the blossom, color may also guide its search for food. Flowers petals often bear obvious dots or lines in a contrasting color. Botanists call these patterns "nectar guides" and believe their purpose is to direct the insect to the flower's nectar and pollen. Even petals that appear unmarked to human beings may carry markings visible to insect visitors. Karl Daumer at the University of Munich first described these "invisible" nectar guides, and Meinwald and Eisner and their research groups at Cornell University later investigated the chemical compounds responsible for them. Black-eyed Susans (*Rudbeckia hirta*) are widespread American wildflowers with deep-yellow, lobeshaped petals. In the center, holding nectar and pollen, is the dark disc that gives the flower its name. To the human eye, the petals appear uniformly yellow and have no obvious markings. Under ultraviolet light, however, the inner portion of the petals is dark, indicating the presence of a chemical compound absorbing the ultraviolet light. This compound adds nothing that humans can see to the petals, but to insects that see ultraviolet light it is a visible pigment. Where humans see only solid yellow petals, a bee sees a nectar guide leading it to the center of the flower, where it wants to be.

Organisms other than flowers also use their colors to promote reproduction. Some advertise themselves openly to attract potential mates. Male guppies (*Poecilia reticulata*) flutter their brightly colored tail fins in front of females during courtship. Their tails come in many different color patterns, and female guppies are particularly enthusiastic about new males that display exotic tail colors. Male birds of many species are brightly colored and spend much time parading before less gaudily attired females. While most species cannot match the extravagant display offered by peacocks (*Pavo cristatus*), many follow the same plan.

To a human observer, one of the more entertaining performing birds is the male blue-footed booby (*Sula nebouxii*), a middle-sized bird common in the Galápogos Islands of the eastern Pacific. The booby struts to and fro in front of his prospective mate, lifting his bright blue feet high in the air and apparently showing them off proudly to the female. She usually remains aloof and unimpressed, perhaps because she has bright blue feet of her own. Male mallard ducks (*Anas platyrhynchos*) have green heads, and females much prefer these males to those whose head color has been artificially altered to a different color.

Color not only attracts receptive females to males, but it may also permit the two sexes to recognize each other. Male yellow-shafted flickers (*Colaptes auratus*) have a black "moustache" marking at the corner of their

6.2 In addition to displaying his high-stepping walk, a male blue-footed booby attracts attention by taking off and circling around a female. On coming in to land, he throws his blue feet up and waves them in front of her.

mouth. A female flicker with a moustache painted on her was attacked by her mate. He accepted her once again when the marking was removed. Eastern fence lizards (*Sceloporus undulatus*) react similarly to altered markings. The male's belly is blue, and the female's is gray. Lizards mistake the sex of those with cross-painted bellies.

ORGANISMS that proclaim their presence in bright colors are not always communicating with their own kind. Many of them want to remind other species that they are dangerous and should be left alone. We mentioned earlier the monarch caterpillar's black, white, and yellow bands and the skunk's characteristic black and white. The caterpillar is poisonous, and the skunk dispenses an evil-smelling spray. These distinctive colors are easily remembered by would-be predators that survive an unpleasant encounter with a caterpillar or skunk. The next time these predators see the striking creature, the memory of their earlier experience may restrain them from repeating it. Because warning colors come from chemical pigments, we looked upon them earlier as a form of chemical defense. We are now looking at these chemicals as lifestyle constituents, and we can appreciate warning colors as one of the several ways living organisms employ their pigmentation.

All these advertisements in color are essentially honest statements broadcast by organisms that want to be noticed. Whatever these creatures' purposes, they present a reasonable version of the truth. However, messages can also be misleading or false. Deception is a way of life for many creatures, and color is often central to the ploy.

Some deceptive organisms are nonpoisonous, palatable creatures that find safety in passing themselves off as toxic. Because predators avoid toxic organisms marked with warning colors, a tasty creature can protect itself by adopting these colors and pretending to be undesirable. Birds refuse poisonous adult monarch butterflies, but they could feed with impunity on nonpoisonous viceroy butterflies (*Limenitis archippus*). Nonetheless, birds also refuse viceroys, because they look very much like monarchs. The two species are similar in size and shape, and viceroys have the familiar orange wings with black veins and borders that make the monarch so distinctive. On the wing, the two species are virtually indistinguishable.

Monarchs and viceroys live together across much of North America, and over most of their wide range the viceroys' deception is successful. In the peninsula of Florida, however, monarch butterflies are rare. There is little protection in looking like a monarch in Florida, because the local birds have never seen one. Here, viceroys masquerade as queen butterflies (*Danaus gillipus*), a local poisonous species closely related to monarchs. Florida birds are familiar with queen butterflies and reject them. Like monarchs, queen butterflies are orange, but they lack the monarchs' bold black wing markings. In Florida, viceroy butterflies have much reduced black veins and borders on their wings and successfully pass themselves off as poisonous queen butterflies.

Many creatures are gifted actors, and color may be only one component of a more elaborate deception. Hover flies are small, innocuous beasts that live on pollen and nectar and serve as important pollinators of flowering plants. Many find safety in disguising themselves as bees or wasps, whose toxic sting keeps potential predators at bay. One deceptive hover fly is *Spilomyia hamifera*, a palatable species that goes through life disguised as a yellow jacket (*Vespula* species). The fly has not only adopted the wasp's dark brown and yellow bands, but carries its mimicry even further. Yellow jackets have long black antennae. Fly antennae are short, so the hover fly imitates a yellow jacket's antennae by bringing its black hind legs forward and waving them in front of its head. The wasp rocks from side to side when it sits on a flower, and the fly mimics this behavior by waving its wings. When disturbed, the fly emits an impressively wasplike buzz.

An imitation can succeed even when there is little biological relationship between model and mimic. In the Kalahari Desert of southern Africa live a beetle (*Anthia* species) and a lizard (*Heliobolus lugubris*) that seem to have nothing in common. The beetle is up to 5 centimeters (2 inches) long, and shiny black with a handsome white lateral stripe. Its eye-catching black and white attire reminds others that it squirts acid on enemies that venture too near. The adult lizard is 10 centimeters (4 inches) long or more. It moves across the sandy desert floor with smooth lateral undulations, well camouflaged by its pale red-tan color.

The connection between these two creatures is that the juvenile lizard passes itself off as a beetle. The young lizard's tail is sand-colored, but its body, unlike the adult's, is black with white markings. It walks with a stiff, jerky gait, its back arched and its tail pressed close to the ground. With its body prominently displayed and its tail nearly invisible, the young lizard resembles the beetle closely enough to deceive human observers. As the juvenile lizard outgrows the beetle, it abandons its disguise and assumes the coloration and more fluid movements of the adult lizard.

A different sort of deception has made cuckoos famous for centuries. Parasitic cuckoos lay their eggs in other birds' nests and then abandon them, leaving foster parents to brood their eggs and feed their young. A newly hatched cuckoo of some species ousts the rightful nestlings or un-hatched eggs and expropriates its foster parents' full attention. In other species, the young cuckoo shares the nest peaceably with the hosts' own young. Birds in several other families, including American brown-headed cowbirds (*Molothrus ater*), follow similar lifestyles.

For cuckoos to reproduce, host birds evidently must accept the foreign cuckoo eggs they find in their nests. Brooding birds can be particular about their eggs, and many species have no mercy on an alien that is recognized. Some birds push a foreign egg out of the nest. Others break the egg and eat its contents. Still others simply abandon their nest and eggs. Cuckoos survive only because they have learned to make their eggs look like those of their hosts. They imitate the shell colors of their hosts' eggs well enough to make their eggs acceptable. Cuckoos that parasitize a single host species have been able to specialize, and they have developed egg coloration to a remarkable level. In southern Europe, great spotted cuckoos (*Clamator glandarius*) parasitize only magpies (*Pica pica*). Magpie eggs are pale green and sprinkled with gray-brown spots and speckles. The cuckoos' eggs reproduce this pattern almost exactly. In India, cuckoos known as koels (*Eudynamis scolopacea*) lay their eggs only in the nests of house crows (*Corvus splendens*). The cuckoos not only match the blue-green color of the crows' eggs but also faithfully reproduce its specific tone, which varies from region to region across the country. Unrecognized cuckoo eggs frequently turn up in amateur collections of bird eggs. Some imitations are good enough to withstand all but the most painstaking scrutiny by professional ornithologists.

LIVING organisms have evolved many other ways of exploiting their coloration, but again and again the basic goals are either making themselves inconspicuous or publicizing their presence. Countless creatures in all corners of the world follow lifestyles where coloration is critical to survival. In these organisms, the same pigments appear repeatedly to convey vital messages in color. Scientists have appreciated the significance of color for more than a century, but field biologists continue to encounter novel examples of its effects.

7

CHEMICAL MESSAGES
WITHIN THE FAMILY

*L*IVING ORGANISMS explore the
world around them in diverse ways.
Various senses permit them to see objects, hear movement, detect magnetic
fields, and receive various other stimuli. Probably none of these senses has a
longer history than the capacity to perceive chemicals. Even a simple bacterium has chemical receptors on its surface and swims toward a bit of sugar
or away from a noxious compound. Here then is another significant role for
chemicals in nature: they carry information. An organism can learn what is
going on in its environment from chemicals that it senses.

Humans sense chemicals primarily by olfaction, that is, by smelling
them. Because the human sense of smell is relatively limited, it is difficult
for us to appreciate the significance odor has for other creatures. We depend largely on sight and sound for our knowledge of the outside world,
and humans who lack a sense of smell often lead quite normal lives. For a
human, being unable to smell is a trivial handicap in comparison with being
unable to see. On the other hand, many other creatures receive most of
their information through their chemical senses. Their sense of smell is
often incredibly keen and may supply highly specific details about their
world. Dogs, for example, are a thousand to a million times more sensitive
to various odors than humans are. They recognize one another individually

by odor, just as we identify our acquaintances by a glance at their faces. A dog that lacked a sense of smell could never lead a normal dog's life.

The ability to sense chemicals and draw information from them is widespread in the natural world. For the organisms that sense them, chemicals carry a variety of meaningful messages. The world is filled with these messages, some passing from one organism to another, others reaching a number of organisms from single events or features in their environment. Creatures downwind flee an approaching fire at the smell of smoke. Seaweed eggs in the ocean release chemicals that attract sperm cells swimming nearby and thus improve their chances of being fertilized. Human egg cells behave similarly. A honey bee emits a chemical signal that recruits other bees to attack the same site at the moment she sinks her sting into a threatening enemy. Snakes track their prey by its scent, flicking their long forked tongues in and out to "taste" the trail as they slither along. A homing salmon returning from the ocean to spawn seeks out the unique scent of its native stream.

Chemical messages raise a host of questions for curious biologists and chemists. What chemicals carry the messages? What connection is there between the structure of a chemical compound and the message it carries? How do organisms distinguish similar chemicals one from the other? How do they recognize complex odors that blend many chemicals together, such as the scent of a stream of water or the smell of a forest fire or the odor of an individual dog? How do they process this information and respond to it?

We are far from having general answers to such questions, although scientists have explored chemical signals for decades. Much of what we have learned comes from observing organisms that communicate with other members of their own species. This includes the honey bee's marking of an enemy for further attack, sperm-attractants broadcast by eggs, and thousands of other messages. These intraspecific signals are the subject of this chapter. In chapter 8 we examine signals passing between members of different species, such as those between a predator and its prey.

Intraspecific signals are rather like a collection of words chosen from several languages. A particular chemical compound may carry totally different messages for unrelated species, just as *camera* means "room" in Italian but something quite different in English. Also, different organisms often use totally different chemicals to carry essentially the same message, just as the article of clothing called a "shirt" in English is a *chemise* in French and a *Hemd* in German.

As these analogies suggest, there is little relationship between the chemical structures of compounds and the messages they carry. Nonetheless, not every chemical can act as a given message. For successful communication, there are practical properties of the chemicals to consider. If the signal is released in the ocean, for example, the compound should not decompose too rapidly in water. A signal that must persist in one place for hours or days, such as one that marks a territory, should not be highly volatile. Limitations such as these are obvious, and in general, all sorts of chemicals serve as signals, from carbon dioxide to elaborate proteins. Most of our present understanding of these signals has come in the past thirty-five years, but scientific awareness of chemical messages goes back more than a century.

THE NIGHT OF THE MOTHS

The first hints that organisms might broadcast chemical messages to their own kind came in the 1870s from startling observations made by French naturalist Jean-Henri Fabre. One day in his workroom, Fabre was examining a female greater emperor moth (*Saturnia pyri*) that had just emerged from her cocoon. With its cream-and-brown wings and a 15-centimeter (6-inch) wingspan, the greater emperor is an altogether handsome animal and the largest moth in Europe. Fabre followed his usual practice and placed the new female under a wire mesh cover for observation.

Nothing exceptional happened during the day, but that evening Fabre's young son came rushing to his father and urged him to come see the moths. Fabre followed the boy to the workroom and found dozens of male greater emperor moths streaming in through the window and fluttering frantically around the wire mesh that confined the female. The males returned for several evenings, pouring through every open window in the house and trying to reach the female moth. The greater emperor was uncommon locally, and Fabre reasoned that the swarm must include males from more than a mile away. He concluded that a signal from the female moth lured them from the countryside to her cage. Fabre suspected that the female emitted an odor that guided the males to her cage, but his efforts to demonstrate this were inconclusive.

141

Fabre's experience with a roomful of moths later became widely known, but contemporary scientists paid little attention to his findings. The scientific community had little regard for Fabre or his work because he lacked the credentials of a professional entomologist. Fabre had completed a doctorate in natural sciences, but he was largely self-taught and had no affiliation with any recognized research institution. At about the time of his moth experiments, moreover, he lost what institutional affiliation he did have. Fabre had spent years as a schoolteacher in the south of France, but his unconventional opinions finally cost him his job. Believing that scientific education should be open to all high-school students, he admitted girls to his science classes. Such liberality was totally contrary to the rules and was regarded as unforgivable. Infuriated superiors responded swiftly. Fabre was denounced from the pulpit and dismissed from his position. Although his extensive popular writings brought him national stature and public recognition late in his long life, the scientific establishment ignored Fabre to the end.

Even though Fabre's work with moths was disregarded, conventional entomologists soon made similar observations of their own. As early as 1882, an American entomologist, Joseph A. Lintner, suggested that female moths might release chemicals to attract males. Half a century would pass before this suggestion was proved correct. For several decades, entomologists studied attraction in many different butterflies and moths. Initial results were puzzling, but clever experiments finally established that the males indeed respond to a scent released by the females. To determine how far the scent carried, scientists set caged females not far from a railway line and released males at different points from moving trains. The surprising outcome was that a male several miles downwind picked up the female's signal and flew to her.

A particularly telling experiment depended on the fact that insects' olfactory receptors (the structures responsible for detecting odors) are in their antennae. Knowing this, entomologists cut off male moths' antennae or covered them with lacquer and then tested these males' attraction to a female. Males deprived of their sense of smell could no longer locate a female. Skeptics proposed that the antennae were really receiving radio signals broadcast by the female rather than odors. Other investigators discredited this interpretation by showing that the sex attractant was present in the air surrounding the female.

Biologists ultimately identified and removed the gland that stores the female's attractant. Males came to the excised gland but not to the female deprived of it. By the late 1930s, the results fit together well enough that most interested scientists agreed on their interpretation: female moths and butterflies release tiny amounts of chemical compounds that attract males. Most species have their own specific signal, to which the males are exceedingly sensitive. Males several miles away respond by flying upwind to reach the female.

THE SILKWORM MOTH'S SEX ATTRACTANT

With the chemical nature of sex attractants established, an obvious challange for chemists was the identification of these chemicals. While biologists now agreed that they were working with chemical signals, no one had any idea what chemicals these might be. In 1939, Adolf Butenandt, a German chemist and recent Nobel laureate, set out to identify a moth attractant. He chose to work with silkworm moths (*Bombyx mori*), which he could obtain commercially. One of Butenandt's most demanding problems was getting enough of the attractant for his studies. At the time of his research, chemists required at least several milligrams (the weight of about 20 grains of salt) of a pure chemical compound in order to identify it with certainty. Unfortunately for Butenandt, female moths make very little of their attractant, because the males are exquisitely sensitive to its call. Butenandt and his collaborators dissected the attractant gland from over half a million female moths to obtain enough material for identification. In 1959, after twenty years of research, Butenandt announced the chemical identity of the silkworm moth's sex attractant. He named the compound "bombykol," combining the moth's generic name, *Bombyx*, with *Alkohol*, the German word for "alcohol," which designates the type of chemical compound bombykol is.

The success with bombykol popularized research on chemical signals. Chemists and biologists quickly identified signals throughout the insect world, and investigation moved rapidly for the next twenty years. Early in the study of insect signals, Peter Karlson and Martin Lüscher coined the term "pheromone" for intraspecific chemical signals, from the Greek words *pherein*, to transfer, and *hormon*, to excite.

Sex attractants turned out to be common among insects. Their pheromones usually consist of one or a few relatively simple chemical compounds.

Improved techniques and equipment permitted chemists to work with smaller and smaller amounts of material with increasing ease. Owing to these advances, chemists can now identify an attractant obtained from just a few insects rather than the half-million female moths that Butenandt dissected. The task now can require as little as a few weeks' work instead of a twenty-year effort.

At the same time, behavioral biologists discovered that male as well as female insects emit chemical signals, and that sex attractants are not their only signals. They found other pheromones in many kinds of insects, including a remarkable variety among ants, termites, and the social bees and wasps.

These four kinds of insects lead communal lives characterized by an extensive division of labor. Individual insects belong to one of several castes and have specific duties, so that no one insect performs more than a fraction of the many activities necessary to sustain the community. Castes such as workers, soldiers, and breeders all have their essential roles in communities that may number thousands or even millions of individuals. Maintaining such complex societies successfully requires extensive internal communication. Pheromones provide much of the communication that regulates these orderly lives.

These social insects have pheromones that direct the recipient's behavior, such as fighting or fleeing, and also ones that alter its physiology. In the honey bees, for example, workers are all females, but their ovaries remain immature owing to a pheromone they receive regularly from their queen. This signal assures that only the queen lays eggs in her own hive. If the queen dies and the pheromone ceases to reach the workers, some of them develop mature ovaries and begin to lay eggs.

LEARNING FROM ANT PHEROMONES

The smooth operation of an ant colony depends upon ten to twenty different signals, and most of these are pheromones. In their monumental work, *The Ants*, renowned authorities Bert Hölldobler and Edward O. Wilson estimate that red imported fire ants employ at least twelve different chemical signals. The simplest of these is the carbon dioxide from the respiration of an ant cluster, which acts as a pheromone to promote aggregation. Workers move toward a source of carbon dioxide, and so solitary ants move to join a group. At the other extreme, the most complex of the fire ants' signals is probably colony odor, by which the workers of a particular colony or nest identify another worker as local or foreign. Each ant nest has its own odor as a result of its location, history, and local food supply. The resident ants pick up this odor on their bodies, so that ants of the same species, but from different nests, have different colony odors. This odor allows the ants to identify intruders and maintain colony integrity.

Fire ants also make use of an alarm pheromone to alert workers to an emergency, and their scouts lay down a trail pheromone to guide the colony during mass migrations. A fire ant queen emits a chemical signal that identifies her to the colony's workers. They respond by scurrying to gather around her. The decomposing corpse of a dead ant also generates its own signal, to which workers respond by eliminating the corpse from their nest.

Ant pheromones also offer biologists opportunities to explore broader questions. With many species and many signals to examine, biologists quickly have progressed beyond cataloguing signals and their functions. One question they have asked is whether the message carried by a pheromone is public or private. Can a pheromone ever have any meaning for species other that its intended recipients? If so, do organisms pay a price when outsiders monitor and decipher their signals?

Ants provide examples of both public and private messages. One of their most important private messages concerns food, for a food source is worth keeping secret from others, including of course other species of ants. Ideally, an ant colony should conceal its food trails to prevent other species from following them to its source of food. In fact, unrelated ants use totally unrelated compounds to guide their own foragers from the nest to food supplies. Each species marks its trails with signals that are meaningless to others, so that an ant crossing a trail left by another ant species typically notices nothing. On the other hand, a secret signal to mark a dead body is unnecessary. Many kinds of ants, as well as honey bees, perceive a natural decomposition product of dead insects as a signal to pick up the corpse and take it from the nest. If an outsider recognizes this message and moves the body, no harm is done.

Through exploring ant pheromones, biologists also have learned that different concentrations of a chemical signal can deliver different messages. One leafcutter ant (*Atta texana*) uses a certain compound as a pheromone to summon help when under attack. At a low concentration, this chemical simply attracts leafcutter ants from afar, but higher concentrations quickly bring the ants to a fighting frenzy. Because the concentration of the pheromone rises as attracted ants approach its source, this one signal both recruits leafcutters to the site of attack, and then, as they arrive, prepares them for combat.

MALE BEETLES PRETENDING TO BE FEMALES

Scientists have studied broader questions about pheromones in other insects, as well. In the 1970s, biologists were surprised to discover that various male insects sometimes emitted the females' sex pheromone. No one offered an explanation for this "mistake" until about ten years later, when Klaus Peschke, then working at the University of Würzburg in Germany, found a justification for it by studying a particular beetle.

To appreciate Peschke's discovery, we must understand how this beetle (*Aleochara curtula*) lives. This little animal passes its life in and near carrion in close association with blow flies, such as the common European bluebottle (*Calliphora vicina*). The beetles devote their entire existence as both larvae and adults to preying on the flies. The blow flies lay their eggs in an animal carcass, and the eggs hatch to maggots (fly larvae) that feed on the carcass and are themselves eaten by the beetles. Maggots that escape the beetles migrate out of the carcass as they mature and then pupate a short distance away. The female beetles eventually leave the carcass as well. Soon after mating, when they are ready to lay their eggs, the female beetles search out the nearby fly pupae and lay their eggs on them. When the eggs hatch, the beetle larvae bore into the fly pupae to feed. Because the female beetles leave to find sites for their eggs, those beetles remaining in a carcass are predominantly males.

Perhaps it is the short supply of females that agitates the male beetles. Whatever the reason, the adult males are combative and spend much of their time fighting among themselves. They bite legs and antennae, beat on one another, and smear defensive chemicals around. The conflict is ceaseless, and the stronger males drive the losers from the habitat, depriving them of food and the opportunity to mate.

Unmated female beetles in the carcass ignore this turmoil. They are protected from male aggression by a pheromone on their cuticles. When an adult male recognizes the female's signal, he refrains from attacking a female beetle but grasps her for mating instead.

The adult males' belligerent lifestyle has one significant drawback. It seriously threatens the lives of young male beetles and so places the entire community in jeopardy. Young males could never defeat older and larger adult males in battle. Once the young males lost to their elders, they would be ejected from their carrion home and would starve to death. No new generation of adult males would develop, and survival of the beetle community would be endangered. The adults males persist in their warlike ways, so the beetles must have developed some way to avoid jeopardizing their own future.

Peschke discovered how the beetles handle this problem. He found that immature male beetles play a chemical trick to avoid being attacked. They ward off the older males by producing and carrying the same cuticular pheromone that female beetles do. Adult males recognize the signal and respond by attempting to mate with them. The young males ignore these homosexual advances from their elders and continue their endless feeding on maggots. The price they pay for safety is being grabbed occasionally by an older beetle. As the young males mature, their protective coating of female pheromone disappears, and they take their place as adult male beetles.

Peschke's observations illustrate that pheromones develop in unexpected and opportunistic ways, as do other aspects of living systems. This need not be the only reason for male insects to appropriate a female pheromone. Other species may meet quite different needs in this way.

HOW DO PHEROMONES WORK?

While scientists were exploring these broader characteristics of pheromones, they also began asking how pheromones work. How does the spectacular swarming of male moths that Fabre first observed come about? A resting female moth emits an attractant that drifts away on a gentle breeze. A male moth downwind twitches his antennae and picks up her signal. It is only one of many airborne chemicals that bombard his antennae, and he must extract the female's message from this sea of chemical noise. How does he do this? How does he then translate the message into the action of flying to a potential mate? As he flies, how does he stay on course to reach his goal? Every sort of pheromone poses comparable questions. There are countless questions but as yet only a few answers.

As for the male moth's flight plan, he apparently stays on course by changing direction when the signal grows faint. The pheromone he detects is spreading downwind from the female in a roughly conical plume that becomes broader and more diffuse as it travels farther from its source. The male begins his flight more or less into the wind and in a straight line. If the signal weakens as he flies along, he turns sharply so it becomes strong again. He maintains this new course until the signal weakens again, and then he makes another turn. The result is a zigzag flight pattern, roughly into the wind within the signal plume, with changes of direction at the edges of the plume. The signal plume narrows and becomes more concentrated as the male moth approaches its source, efficiently guiding him to the waiting female.

Before his flight began, the male moth received the female's chemical call and interpreted it in his brain. Several research groups are working to trace the series of events that begins with arrival of the pheromone at the male's antennae. The overall problem is to analyze how the brain translates the signals generated in the antennae into flight to the female. This is an enormous research undertaking, and neuroscientists are only beginning to understand details of the earliest steps. One important contributor here is a group at the University of Arizona led by John G. Hildebrand.

The Arizona scientists are tracing the fate of attractant signals in the sphinx moth (the adult form of the tobacco hornworm, *Manduca sexta*). The female moth releases an attractant that is a mixture of several chemical compounds, but only two particular components are necessary for the male to read her message. These two compounds are chemically similar to each other, and we can refer to them simply as the major and minor components.

When the gentle breeze brings these two components to a male moth's antennae, they come in contact with protein olfactory receptors. The antennae carry several kinds of receptors, each one possessing a unique structure and molecular shape. This arrangement permits each receptor to interact with particular types of chemical compound. One kind of receptor on the moth's antennae responds to the major component of the attractant, and another responds to the minor component. The receptors that respond to the major component do not recognize the minor one, and vice versa, despite the two components' chemical similarity. The molecules of each component have a shape that allows them to fit snugly only into their own receptor. When the receptors interact with their appropriate compounds, they send nerve impulses to the male moth's brain.

Nerve impulses from all the receptors in the antennae go directly to a part of the moth's brain called the antennal lobes. The moth has two antennal lobes, one for impulses from each antenna. In male moths, each antennal lobe is divided into three separate structures. Hildebrand has discovered that receptors for the major component of the attractant send impulses exclusively to one of these three structures, while receptors for the minor component send their impulses to another one of the three. Nerve impulses from other kinds of receptors in the male antennae pass to the remaining third structure. (The female moth's brain lacks the two structures devoted to impulses generated by the attractant.) In the male's antennal lobes, perhaps 5 to 10 percent of all the neurons are involved in processing signals generated by the female sex pheromone.

The antennal lobes relay these signals from the antennae to other parts of the moth brain. When the two structures dedicated to the major and minor components of the attractant are both briskly passing signals on, the brain interprets this pair of signals as the female's attractant pheromone. The brain then responds by generating nerve impulses of its own, but the many steps that follow are still obscure. Ultimately, these new signals lead to the moth's spreading its wings and flying into the wind.

FIGHTING PESTS WITH PHEROMONES

Particularly in the early years of research, there was widespread hope that insect pheromones could replace pesticides in controlling agricultural and forest pests. The concept is appealing in its simplicity. Scientists can identify the sex attractants of insect pests. Chemists can synthesize these attractants in the laboratory, and engineers can manufacture them on a large scale. Farmers should be able to control specific crop pests using these synthetic pheromones rather than by spraying fields with indiscriminately lethal pesticides. Scientists labored to fulfill this hope, and for several years funds were available to support their efforts. In practice, things turned out to be much more complicated than expected.

In the 1970s, two strategies emerged for the practical application of pheromones to control insect pests. In the "trap-out" strategy, farmers and foresters distribute traps baited with a pest's attractant pheromone throughout their fields and forests. The chemical message spreads from the traps, attracting insects from near and far. Pests that come seeking a mate are caught in the traps, or "trapped-out." In a modification of this procedure, the traps contain both a sex attractant and a chemical insecticide. When the insects arrive, the "attracticide" trap kills them. The attracticide technique does not forgo pesticides completely, but it confines them to the traps and requires much smaller quantities than ordinary spraying. It also selectively destroys only the pest species lured into the trap, leaving beneficial insects unharmed.

An alternative control strategy called "mating disruption" is particularly suited to controlling the many crop pests that are not adult insects but larvae. This strategy calls for spraying entire fields with a synthetic version of the insects' attractant pheromone. The false signal floods the area, and, among other effects, it overwhelms the signal emitted by individual insects. The natural signal is lost in this chemical deluge, and no clear odor trails remain to guide an insect to a receptive partner. A female broadcasts her attractant, but no male can find her. As a consequence, male and female do not connect, and their crop-eating larvae are never produced.

Either the trap-out or the mating disruption strategy should break the targeted pests' life cycle. The pest species should be destroyed or at least held in check, while beneficial insects remain untouched. Additionally, because insects are so sensitive to their attractants, both strategies require relatively small amounts of chemicals. Furthermore, unlike ordinary pesticides, the synthetic pheromones are not toxic. As a result, these strategies avoid destructive pollution.

Practical field trials of each of these strategies took place about 1970. Tests of the trap-out strategy against an infestation of bark beetles in the conifer forests of Norway and Sweden gave somewhat equivocal results. Thousands of millions of beetles were destroyed, but nonetheless, damage to the forests was extensive. Whether the damage would have been even worse without the pheromone traps is debatable.

A large trial of the mating-disruption technique against an American cotton pest gave more clearly positive results. This trial covered an entire growing season, as governmental agencies, scientists, and cotton growers joined to combat pink bollworm. This "worm" is the larva of a small brown moth, *Pectinophora gossypiella*. It is one of the world's most destructive pests and plagues cotton (*Gossypium hirsutum*) fields from the East Coast to California.

To test mating disruption as a technique to control pink bollworm, agricultural scientists enlisted the cooperation of cotton growers in Southern California. The scientists devised a spraying protocol that all the growers throughout the Imperial Valley agreed to follow for an entire growing season. At the end of the season, the participants agreed that mating disruption had effectively controlled pink bollworm in the trial area. Because beneficial insects had been spared, other pests had remained under control as well. The direct cost of the trial had been competitive with the cost of conventional spraying.

While the test was a technical success, it also underscored the complexity underlying the practical problem of pest control. The Southern California experiment precipitated bureaucratic and political objections, brought industrial resistance to replacing pesticides with pheromones, and revealed economic problems. In spite of the victory in the field, these practical difficulties have deterred subsequent large-scale use of mating disruption against major agricultural pests.

Nevertheless, limited specific applications of pheromones have proved more generally successful. One of these is a mating-disruption approach to fighting a sugarcane (*Saccharum officinarum*) pest in the lower Rio Grande Valley of Texas, which was invaded by the Mexican rice borer (*Eoreuma loftini*) in the 1980s. This insect soon became a major problem in the area and was destroying 18 to 20 percent of the local sugarcane crop. USDA scientists stepped in with a synthetic version of the female rice borer's sex attractant as a mating disruptant, and crop damage dropped to 3 to 4 percent. Instead of spraying the attractant, workers spread small pheromone-scented plastic or rubber chips throughout the cane fields. The chips remain active sources of the pheromone for more than 140 days, so it is unnecessary to treat the fields again and again.

In other efforts, agricultural scientists have developed and tested promising mating-disruption techniques for a variety of pests. In 1993, the Environmental Protection Agency (EPA) approved a new formulation to combat the tomato pinworm (*Keiferia lycopersicella*), a pest that has been a problem for years. Controlling tomato pinworm economically with conventional pesticides has been difficult, and pesticide residues on tomatoes have been a continuing menace. The pheromone approach may provide a solution to both problems.

Despite these achievements with insect sex pheromones, their most widespread practical application today is in monitoring insect populations. Traps baited with a particular pest's attractant are set out, and from the numbers of insects captured, entomologists calculate the total population of the species present in the trapping area. Farmers deploy traps to estimate insect populations before and after spraying and so avoid applying insecticides excessively. Commercially available traps placed in citrus groves warn of the arrival of occasional pests before they cause grave damage. Networks of traps alert governmental agencies to incipient infestations of several exotic pests, such as the Mediterranean fruit fly, or medfly (*Ceratitis capitata*). Along the southern borders of the United States, there are forty thousand traps baited with the medfly's attractant pheromone. These traps serve as an early warning system in an ongoing effort to keep medflies out of the United States.

A fundamentally different strategy for pheromonal control of pests that has appeared recently holds promise for gardening as well as agriculture. Unlike previous schemes, this strategy depends on attracting beneficial insects rather than pests. The idea is to recruit a hungry insect-eating species to annihilate the destructive pests. In a strategy developed at the USDA, the hungry predator being actively recruited is the spined soldier bug (*Podisus maculiventris*), a small brown creature that feeds on the eggs and larvae of several other insects. Like other species that entomologists call "true bugs," soldier bugs have mouthparts designed for piercing and sucking. Many true bugs use these structures to suck up plant juices, but soldier bugs are carnivores. They stab a caterpillar with their barbed, needlelike beaks, inject enzymes that digest and dissolve their victim's body from within, and then feed on the juices.

Soldier bugs are willing to attack a variety of insects, making them ideal protectors of crop and garden plants. Their diet includes the larvae of gypsy moths (*Lymantria dispar*), Colorado potato beetles, cabbage loopers, and other common pests. When a male bug finds a promising supply of these

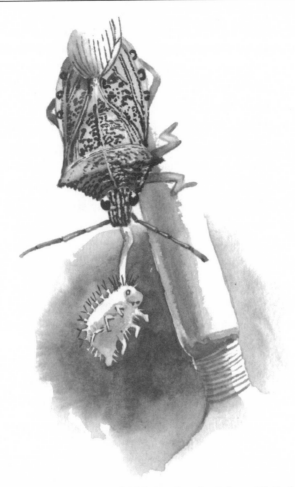

7.1 *A soldier bug feeds on an impaled Mexican bean-beetle larva* (Epilachna varivestis), *which is a serious pest on all kinds of cultivated beans. Adult soldier bugs emerge from the ground in the early spring, looking for food and mates just in time to rid the garden of harmful insects.*

delicacies, he broadcasts a chemical message that brings other soldier bugs to share his discovery. Both males and females come and feed voraciously. One entomologist followed a single soldier bug's feeding for nine weeks and found that in this period it consumed 122 beet armyworms, about two armyworms a day. (Soldier bugs are about 1 centimeter [0.4 inch] long, while armyworms grow up to three times that size.) During their feeding orgies, soldier bugs also take time to reproduce. When a female answers the male's call, she arrives ready to mate. She can lay about one thousand eggs over five to eight weeks and so contribute an army of new soldier bugs to the area.

Owing to the USDA's research, we know what chemical signal the soldier bug releases. A team of government scientists led by Jeffery R. Aldrich investigated the pheromone and found that it is a mixture of two previously known natural chemicals known as leaf aldehyde and alpha-terpineol. Leaf aldehyde is in the sweet fragrance of freshly mown grass, while alpha-terpineol is a component of pine oil. As a result of this research, the USDA has patented a synthetic version of the soldier bug attractant. The synthetic pheromone is now privately manufactured and marketed under the trade name Rescue. Dispensers of this synthetic attractant are set out in fields and garden plots, and bring hungry soldier bugs to feed on unwanted pests.

For gardeners, the soldier bug strategy may offer real advantages over the older trap-out technique. Baiting a few traps with a pest's attractant can bring more pests to an area than the traps can accommodate. Long after the traps have filled with insects, they continue to release the irresistible pheromone that encourages more pests to hasten to the garden. Those that arrive after the traps are full simply turn their attention to their natural activity, stripping the garden. In this way, too small a number of baited traps can be worse than no traps at all. The new soldier bug strategy employs no traps; every bug drawn to the garden devours pests, not plants.

Other ways of combatting pests with pheromones appear as scientists decipher new chemical messages. One unproved possibility is to deploy pheromones that deter egg-laying. The females of at least thirty-three insect pests leave a chemical signal to mark the site where they have laid their eggs. This signal deters other females of the same species from laying eggs in the marked site. A second female seeking a site for her eggs heeds the message and moves on to deposit her eggs elsewhere. This benefits both females' larvae by reducing competition among them for space and food. Distributing larvae over more sites should improve their chances for survival.

A pest that safeguards its eggs in this way is the cherry fruit fly (*Rhagoletis cerasi*). Each female fly lays her eggs singly in half-ripe cherries (*Prunus avium* and related species). She deposits an egg by means of a needlelike tube called an ovipositor that she inserts deep into the fruit. As she withdraws her ovipositor after depositing the egg, she drags the end of it over the surface of the cherry to leave her chemical mark. Other female flies detect this mark, bypass the marked cherry, and lay their own eggs in other fruit. Consequently, cherries containing two fly eggs are rare.

Unfortunately, however, one egg is sufficient to ruin a cherry for the market. Spraying cherries as they ripen with a synthetic version of the deterrent pheromone should protect them from fruit flies. The pheromone would signal female flies that all the fruit in the orchard was already in use as egg sites. They would fly on to lay their eggs elsewhere. Field tests suggest that the idea may be practical.

THE SEARCH for pheromones that began with insects soon extended in all directions. Wherever scientists have looked, they have found creatures using chemical signals. Bacteria, mites, and ticks send messages that bring individuals together in aggregates. Barnacles and crabs use signals between a mother and the eggs she is carrying to ensure that all the eggs hatch at about the same time. Pheromones control the courtship and mating rituals of mites, salamanders, and many fishes. Injured earthworms and termites emit alarms that send others of their species fleeing. House-mouse urine contains signals that accelerate or retard sexual maturation of young females.

Innumerable chemical messages pass back and forth in living communities, shaping the lives of creatures everywhere. They offer one of the most striking manifestations of nature's chemicals uniting the living world. We look at a few of these varied communications.

COOPERATING TO MAKE A PHEROMONE

Typically, one organism makes and disseminates a pheromone, and another receives and acts on its message. However, not all chemical signals work this way. Some simple fungi of the order Mucorales send and receive a mating pheromone, but the senders are also the receivers, and vice versa.

The best-known member of the class Mucorales is probably *Rhizopus stolonifer*, the common mold that forms white tufts on bread. These molds exist in two mating types, designated (+) and (–), that can join in a primitive form of sexual reproduction. In several species of Mucorales, mating is under the control of a chemical signal known as trisporin. To understand trisporin's function, we first need a brief description of the sex life of these organisms.

Each mating type of these molds grows independently as masses of slender tubes or threads. Different species grow on different kinds of organic matter, ranging from horse dung to plants to bread. When growing masses of the two mating types of a particular mold encounter each other, they send up erect tubules called zygophores along their line of contact. The zygophores are structures that allow the two mating types to meet. The erect (+) zygophores stand along their side of the frontier, and the (−) ones along theirs. The ends of the zygophores grow a few millimeters toward each other, arching gracefully over the frontier and in the direction of the other side. The growing tips of a (+) and a (−) zygophore meet and fuse. At the point of fusion, a new round structure develops and enlarges into a body known as a zygospore. This zygospore grows further, germinates, and finally generates new individuals. These new individuals possess genetic material from each of the organisms whose zygophores joined to create them.

For these mating events to commence, an initial message from the pheromone trisporin is essential. Only when it is present do (+) and (−) mating types send up zygophores that grow together and exchange genetic information. Trisporin induces each mating type to generate its zygophores, but neither type can make trisporin alone. The two mating types can synthesize the pheromone only when both types are present.

Synthesizing chemical compounds is a stepwise process. Organisms convert one compound into another by a series of relatively small chemical changes, one after the other. Each step leads to a new compound that differs only slightly from the one before. (Chemists prepare compounds in the laboratory in the same stepwise way. Depending on how great the difference is between the starting compound and the final product, the number of steps needed may be only a few or perhaps more than two dozen.)

Both the (+) and (−) mating types of the Mucorales molds perform the early steps leading to trisporin equally well, but each type can proceed only so far. There is one step leading to trisporin that the (+) type cannot perform, and another step that is impossible for the (−) type. Each type has no trouble carrying out the one step that the other type cannot accomplish. While neither type can complete the synthesis alone, they can produce trisporin by working together.

Each type carries out the early steps leading to trisporin on a continuing basis, releasing the products of its efforts into the environment. Where (+) and (–) types grow into close contact, these released intermediate compounds diffuse across the line of contact from the mass of one type to the other. Each type takes up intermediates from its environment, carries out whatever chemical steps it can along the route to trisporin, and releases the products again. The products of these synthetic efforts pass back and forth, and the two mating types finally produce trisporin. Once they achieve the synthesis and the pheromone is present, it stimulates each mating type to send up its zygophores.

In trisporin, these molds have a simple trigger for generating zygophores and initiating sexual reproduction.

COMMUNICATION AMONG PLANTS

While pheromones have been familiar in molds and algae for years, there was little thought that higher plants might communicate with one another until the early 1980s. Recently, however, the existence of chemical communication between plants has become an exciting possibility. Some disagreement continues among plant scientists about the reality of plant pheromones, and as yet there have been virtually no chemical investigations here. Nonetheless, the evidence for chemical signals in plants is now respectable, and it is rapidly improving.

You may recall that we mentioned a probable plant pheromone in discussing the warfare between lima beans and spider mites. When spider mites infest a bean plant, the plant emits a volatile distress signal. Nearby healthy bean plants apparently respond to this signal by broadcasting the same alarm themselves. If healthy plants do indeed begin sending the alarm as a consequence of an infested plant's signal, communication from one plant to another is taking place. (Carnivorous mites answer the distress signal and save the infested plant by devouring the spider mites. We could have just as easily discussed the lima bean signals here, in terms of pheromones, or in chapter 8, when we explore interspecific signals.)

Lima beans are not the only plants to behave this way. The same spider mites that attack lima beans also feed on cotton, and the carnivorous mites once again come to the rescue. Studies on cotton are less complete, but infested cotton plants seem to send a signal akin to the lima bean's alarm. Thereafter, nearby healthy cotton plants resist infestation from marauding spider mites, and they also attract protective carnivorous mites to their own leaves.

Experiments on cotton at the USDA's Southern Regional Research Center in New Orleans focussed on the defensive chemicals produced by cotton leaves. When cotton is under attack, it fights back by expanding production of defensive chemicals and other volatile compounds. USDA biologist Hampden J. Zeringue Jr. exposed detached but undamaged cotton leaves to volatile chemicals released by other cotton leaves infected with *Aspergillus flavus* (the common fungus responsible for turkey "X" disease). After a week's exposure to these volatiles and whatever messages they bore, the undamaged leaves had increased their own output of defensive chemicals by 34 percent. The healthy leaves had strengthened their own defenses in response to a fungal infection in neighboring but physically separate leaves. The only linkage between the two sets of leaves was a current of air.

Poplar trees (*Populus × euroamericana*) respond in a similar way to distressed neighbors of their own species. After two leaves of small poplars were torn in half, the plants' chemical defenses increased—the typical response many plants make to such abuse. Remarkably, undamaged poplars housed with the injured plants also began to synthesize more defensive compounds in response to their neighbors' injuries. All the plants were individually potted, so the healthy plants were apparently reacting to an airborne message.

Parallel experiments with sugar maple seedlings (*Acer saccharum*) gave the same results. Similarly, healthy Sitka willow trees (*Salix sitchensis*) build up their chemical defenses in response to an assault of tent caterpillars (*Malacosoma californicum pluviale*) on nearby willows.

All these experiments suggest that injured plants communicate their distress to others in their community. In animals, such communication is the business of alarm pheromones, signals that limit predation and reduce damage by forewarning an entire community to take appropriate defensive measures. If plants also employ alarm pheromones, agricultural scientists will be seriously interested in enlisting these chemicals in the war against herbivorous pests. Perhaps a farmer could enhance natural plant defenses by spraying his fields with a crop plant's own alarm pheromone. This might increase crop resistance to pests at very little environmental cost. Investigations over the next few years should reveal whether this concept has practical merit.

CUTTING UP A SPIDER WEB

Chemical signals are much better established in spiders than in plants, but we still know surprisingly little about them. A spider sex pheromone was first identified chemically only in 1993. As you might expect, spiders have combined pheromones with silk in useful ways. One of silk's advantages is that it offers a convenient physical support for disseminating chemicals. The Sierra dome spider (*Linyphia litigiosa*) from the western United States and its common European relative, *Linyphia triangularis*, disseminate pheromones from their silk in just this way. When a female *Linyphia* is ready to mate, but finds no males in sight, she spins herself a new web and adds an attractant pheromone to her silk. The scented web acts as a reservoir that provides an enduring source of the attractant. The signal slowly evaporates and broadcasts its compelling call to male spiders.

7.2 Linyphia triangularis *spiders spin a convex, horizontal sheet-web. A courting male spider cuts the support threads, freeing much of this delicate sheet, and then rolls it up.*

A prospective mate picks up the call downwind and hurries to the female's web. Despite his haste, when he arrives, he does not immediately approach the female. Instead, he moves actively about her web, snipping many of its threads. He cuts a large part of the web free and rolls it up into a small ball. A male *Linyphia* spider behaves in this curious manner only on the web of an unmated female. His purpose is apparently to protect his relations with her. By removing a large portion of her web, the male minimizes further evaporation of her sex pheromone. This should reduce the likelihood of attracting a rival spider that would interrupt the male's anticipated liaison.

After attending to possible competition, the male turns his attention to the waiting female. Courtship and mating are lengthy affairs among these spiders. They can last up to five hours, and if a rival male should arrive during this time, the first male must quit the female and fight for his right to retain her. After a male has mated with the female spider, the attractant has served its purpose. Hereafter, the female no longer adds it to her silk.

COURTING SALAMANDERS

The rather primitive, tailed amphibians known as salamanders are another group of animals that offer a host of unexplored problems to pheromone scientists. Salamanders are the living animals most closely related to those venturesome vertebrates that first left the water to spend part of their lives on land. Most salamanders are small and secretive, and so escape notice, but, surprisingly, they are the most common land vertebrates in much of the eastern United States. They generally remain near water or in the moist places their way of life requires, feeding on insects and other small invertebrates, and producing larvae that begin life in the water, as tadpoles do.

There are more than three hundred species of salamanders, and they have widely differing mating habits. The species that interest us here mate on land after complex courtship rituals during which the male passes pheromones to the female. These rituals have been familiar to zoologists since the nineteenth century, and at least one investigator concluded as early as 1927 that they depend on chemical signals. The idea was apparently too far ahead of its time, however, and no one paid further attention for decades. Biologists revived their interest in salamander courtship pheromones only in the 1980s. The chemical nature of these unusual signals is still unknown.

The salamanders' rituals are fascinating. Courtship begins with two salamanders making physical contact. Typically, the male crawls over or under the female, perhaps grasping her with his legs or tail, in a behavior pattern characteristic of his particular species. Eventually, he moves into a position with his head near hers. He rubs her snout with specialized pheromone glands. In some species these are located on his cheeks, in others, under his jaw or elsewhere on his head. Only males have these glands. His rubbing transfers their secretions almost directly to the female's olfactory system.

Male salamanders of some species also possess specialized long teeth that females lack. During courtship, the male scratches the female with these teeth or even bites her. Then he wipes his pheromone glands over the wound he has created. This must inject his courtship signals into the female circulatory system, but the effect of this transfer remains obscure. Depending on the species, two salamanders may pursue their courtship for only a few minutes or perhaps prolong it for more than an hour. When the female is willing to proceed further, the male ceases to ply her with chemicals. The pair then moves on to the next stage of mating.

These rituals and their pheromones are still a puzzle, but biologists have several possible explanations. One derives from a crucial feature of these salamanders' sex life. In mating, the male deposits his sperm in a tidy packet somewhere outside the female's body, on the ground or on a twig, for example. If he is to have any chance of fathering the female's offspring, she must pick up his sperm packet and store it in her body. She will lay her eggs later, fertilizing them just beforehand with the sperm she has acquired earlier. If the female does not take up the male's sperm packet, his mating effort has failed. By then it is too late, and he has wasted considerable energy. Courtship and the accompanying pheromones may serve primarily to heighten the female's receptivity and stimulate her to accept the male's sperm.

MAMMALIAN COMPLICATIONS: VOMERONASAL ORGANS, COMPLEX ODORS, AND LARGE BRAINS

Vertebrates beyond the salamanders have pheromones as well. Turtles and lizards probably use chemical signals, and there are some well-studied snake mating pheromones. There may be pheromones among the birds, although

161

there has been very little exploration of the possibility. It is when we reach the mammals that we find a treasure trove of chemical signals. Mammals have pheromones to mark territories, serve as sex attractants, signal alarm, and carry all sorts of other messages. After the insects, we have learned more about pheromones in mammals than in any other group of creatures.

Pheromones become more difficult to study as we move from amphibians to mammals. One complicating factor is a sense organ that assumes particular importance in mammals. Most vertebrates other than fishes possess a structure called a vomeronasal organ (VNO). This is typically a small, paired finger or pouch located off the nasal cavity or above the roof of the mouth. It is lined with sensory cells bearing protein molecules that are receptors for chemical compounds. These vomeronasal sensory cells look somewhat different from ordinary olfactory cells. They also connect to different areas of the brain. These areas, known as the accessory olfactory bulbs, are separate from the olfactory bulbs that receive connections from olfactory cells. Here then is an independent means of sensing chemicals in the environment, literally a sixth sense, that is unique to vertebrates. This sixth sense is generally known as the accessory olfactory system.

In mammals, the VNO is not an optional second sensory pathway or some sort of backup organ for the olfactory system. It is rather the required means of receiving particular pheromonal messages. Many mammals pick up molecules destined for the VNO when they sniff and lick. In addition to passing into the nose and the olfactory system on a stream of air, many of these molecules are absorbed on the damp surfaces of the nose and tongue. From there, they are transported in saliva into the VNO. In many creatures, this pouch can contract and expand, and so act as a pump to force the liquid that transports chemical signals in and out. Because compounds reach the VNO in liquid, the accessory olfactory system can sense molecules that are not sufficiently volatile to become airborne. In this way, quite large molecules acquired only on direct physical contact may serve as mammalian pheromones. This, then, is an exception to the general observation that pheromones are airborne chemicals.

When dogs and cats lick objects and then lick their noses, they are quickly moving chemical compounds from the environment into their vomeronasal organs. An elephant accomplishes the same thing by touching objects with the tip of its trunk and then placing its trunk in its mouth.

Chemicals reach the elephant's large VNO by way of ducts in the roof of the mouth.

This accessory olfactory system in mammals seems to be devoted primarily to signals of a social nature that affect fundamental aspects of a mammal's life. Some of these pheromones concern attraction and courtship. Others influence reproductive behavior and aggression, or modify the course of puberty, pregnancy, or the female cycle.

The VNO is an interesting complication, but it is not the only obstacle to studying mammalian pheromones. Mammals present several other novel difficulties. They have large brains that integrate information from their various senses. They lead complex lives, and their pheromones may carry correspondingly complex messages. A mammal that receives a pheromonal message may or may not respond in any obvious or measurable way. It may have learned something that requires no response. Perhaps it has filed the information away for future reference. Black-tailed deer (*Odocoileus hemionus columbianus*) sniff and lick tufts of hair on one another's hind legs with obvious interest. The mixture of compounds on the tufts may convey many messages, but no response of the deer discloses what these might be. We know what chemical attracts the deer to the tufts and stimulates their sniffing, but the messages they pick up there remain undeciphered.

Another complication is that mammals have multiple pathways for the same or complementary information. There may be visual or aural signals that repeat or supplement a pheromone's message. When a black-tailed deer is frightened, it may release an alarm pheromone, stamp its feet, erect its tail, hiss, cock its ears, walk stiff-legged, or do some number of these things more or less simultaneously. Other deer take its behavior as a warning of danger. How does an investigator decide to what extent these deer are responding specifically to the alarm pheromone?

Large brains also allow mammals to make extensive use of chemical signals based on complex odors, that is, scents consisting of many different chemical components. Signals based on complex odors are not limited to mammals, but mammals use them widely, particularly for individual and group recognition. A doe recognizes her fawn, and a ewe her lamb, by its own particular odor. A mother elephant seal carefully sniffs her pup just after its birth, and the pup sniffs its mother, each becoming familiar with the other's odor for future recognition.

163

Group recognition is important for mammals that control and defend a particular territory. These animals live in families and strive to prevent other groups of their own species from encroaching on their territory. One such territorial mammal is the klipspringer (*Oreotragus oreotragus*), a small, dog-sized antelope native to rocky areas of the African savannah. Like many territorial mammals, klipspringers depend on pheromones for group identification.

A family of klipspringers, consisting of a mated pair and their young, maintains an exclusive territory. As the group browses or grazes, one member of the family acts as a lookout. This sentinel may chew its cud, but otherwise it stands rigid and attentive on a mound or raised rock. It watches not only for predators but also for other klipspringers that might invade the family's territory. Klipspringers in zoos retain this habit of posting a lookout, so that one of these little antelopes is usually found on watch, erect and motionless on a high point in their space while other family members lounge about.

Klipspringers mark their territory with a chemical signal secreted from a gland on each side of their nose, just in front of the eyes. By rubbing the gland against twigs and bushes, the animal leaves a signal indicating its claim on the area. Other klipspringers recognize the mark as belonging either to a member of their own family or to an outsider. The importance of these marks is clear from experiments carried out with two families of klipspringers at zoos in Frankfurt-on-the-Main in Germany and Naples, Italy.

Biologists in Naples obtained the pheromone-containing secretion from one of the Frankfurt Zoo's animals and spread it about in their own klipspringers' area. The Naples male became nervous and mounted guard as soon as he detected the foreign pheromone in his space. The biologists applied more of the Frankfurt secretion, and the Naples male's agitation visibly increased. He remained on guard and refused to interrupt his vigil, even to eat. The male klipspringer maintained his lookout until the biologists removed the Frankfurt scent from his space. Once the alien signal and the ostensible threat to his family's territory were gone, the male relaxed, returned to his normal habits, and began to eat again.

Odors associated with groups or individuals presumably arise from slight differences in diet or habitat, and small variations in metabolism that modify breath, sweat, and other secretions. Recognition pheromones based

164

on complex odors are comparable to the aroma of coffee or the bouquet of wine. We immediately recognize these scents, but there is no unchanging chemical formula for an odor that says "coffee" or "wine" to us. Typical American breakfast coffee smells unlike Italian espresso, but we instantly identify each as coffee. Such scents vary endlessly through minute changes in a composition that embraces hundreds of components.

Chemically complex signals associated with individuals or families of organisms are obviously different from single-component pheromones used by an entire species. Every female silkworm moth sends out bombykol to attract a mate, but each calf in a herd of cattle has an odor all its own. Because mammalian recognition signals have special characteristics, some scientists prefer to call them social odors rather than pheromones. We cannot yet analyze these complex odors chemically, nor can we explain in detail how organisms interpret them.

Owing to the complexities surrounding mammalian chemical signals, our understanding of them is still quite limited. The signals used by dogs (*Canis familiaris*) provide a telling example. Male dogs mark trees and posts with a spurt of urine that other dogs investigate eagerly and sometimes mark over. Dogs sniff each other in ceremonies of greeting and identification. Males collect in a quarrelsome pack around a bitch in heat, enticed and distracted by her airborne attractant. These are everyday canine activities, familiar the world over, and yet they are a mystery to us. We have not deciphered the messages in urine marks, and we remain uncertain why a dog may cover an earlier mark with his own urine. The chemicals responsible for all of these signals are unknown to us. We know little more about these canine practices today than when humans and wolves first joined forces some twelve thousand years ago.

This is not to say that we are totally ignorant of mammalian chemical signals. There have been several successes in seeking to understand them. Pheromones in the urine of house mice (*Mus musculus*) control female puberty and pregnancy. Their biological and behavioral effects have received close attention for decades. Syrian golden hamsters (*Mesocricetus auratus*) are peculiarly dependent on chemical signals for normal sexual behavior. We have chemical and biological details of the pheromones that attract a male hamster to a receptive female and prompt him to mount her. A boar pheromone that improves sow receptivity is sold to pig farmers as Boar Mate to facilitate artificial insemination. All in all, we know quite a bit about behavior mediated by mammalian pheromones but little about the chemical compounds involved.

WHAT ABOUT HUMAN PHEROMONES?

For most nonscientists, interest in chemical signals naturally concerns the possibility of human pheromones. Do we broadcast pheromones? If so, what kind of messages do we send and receive? What if these signals include irresistible human sex attractants and aphrodisiacs? Regrettably, there is no reason to believe that such delights are in the offing. After all, humans who lack a sense of smell (or sight or hearing, for that matter) can have normal sex lives. Human sexual response is too complex to fall under the exclusive control of any single sensory imput. However, other types of human pheromones are a different matter. There is convincing evidence that humans do use pheromones. One of the best investigated of these signals is common enough that many women have personally experienced its effects.

This signal is responsible for what is sometimes called "women's dormitory syndrome." When women live in groups, their menstrual cycles often become synchronized. A group of women whose cycles are initially scattered within the month may find after a few months of living together that their cycles nearly coincide. In the 1970s, Martha K. McClintock, then working at Harvard University, studied this shift of cycles in dormitory residents at a nearby women's college. She found that the change in menstrual timing depended only on association. The more hours a day two women spent together, the more likely that their cycles would coincide.

If association leads to menstrual synchrony, a chemical signal passing from one woman to another could be responsible. This idea was discussed for more than a decade, until a research group led by George Preti in Philadelphia at the Monell Chemical Senses Center and the nearby University of Pennsylvania located the pheromone that causes menstrual synchrony in women's axillary (underarm) sweat. The sweat changes in composition through the course of a woman's cycle, and over time this changing signal influences the timing of other women's cycles.

Working from another of McClintock's observations, the Monell group also discovered a second pheromone. This is a signal in male sweat, and it imposes regularity on the menstrual cycle. After women volunteers with unusually short or long cycles had repeatedly sniffed an extract of men's axillary sweat over about three months, their cycles were more regular and approached the normal length of twenty-nine days. The cycles of a control group of women who sniffed only ordinary alcohol were unaffected.

These curious pheromones serve no obvious purpose in the lives of modern women. However, female baboons (*Papio cynocephalus*) living together also seem to synchronize their cycles, and Jeanne Altmann, a University of Chicago primatologist, has proposed that this gives them an advantage in choosing mates. The sexual receptivity of female baboons, unlike women, is tied to their menstrual cycle. If an entire group of females becomes receptive at the same time, it is improbable that a single dominant male baboon could block competing suitors from approaching all of them. Consequently, the females should be able to make a choice among prospective mates and avoid falling under the control of a single male.

In the decades since McClintock's observations, popular awareness of chemical communication has grown enormously. In 1975, some people might have heard of sex attractants in butterflies, but the word *pheromone* was still unfamiliar. Nowadays, newspapers discuss the subject quite casually. In the spring of 1992, the press reported that a pheromone "secreted from men's armpits and groins" is the active ingredient in a product being marketed to bill-collectors. This narrow market was targeted, it seems, because an Australian study had revealed that invoices impregnated with the pheromone had a 117 percent higher return than untreated ones. In another story, the "Life & Times" section of the *Times* of London disclosed that a chemist had "almost" isolated a human aphrodisiac pheromone, for which a vast public demand was foretold.

The idea of chemical signals as aphrodisiacs understandably gladdens perfume manufacturers as well. A 1992 advertisement promoted a fragrance with the claim that "a little knowledge of biology can improve your chemistry." There is also a perfume called simply *Pheromone*; according to a full-page spread in a national magazine, "In a word, it means attraction."

Some of the more spectacular and titillating claims made for human pheromones may not survive closer examination, but continuing news reports on the human VNO, pheromones, and perfumes have been sufficiently startling to sustain broad interest. One of the most remarkable of these reports first reached public notice in 1991. Until then, few scientists believed that human beings had a functional accessory olfactory system. The VNO made a supposedly transitory appearance during prenatal human development, and some scientists suggested that it might be important in a newborn infant's recognition of its mother. Most regarded it as vestigial or absent in adults.

It was something of a surprise then, when three research groups announced that human vomeronasal organs were as common as eyes and ears. All three groups presented their results at an international symposium held to discuss research on mammalian pheromones. This meeting took place in Paris in 1991, and its proceedings later appeared in a respected scientific journal. Groups led by Larry J. Stensaas of the University of Utah School of Medicine and David T. Moran of the University of Colorado School of Medicine reported that they had independently examined the noses of hundreds of human beings and found that virtually all of them contained paired bilateral vomeronasal pits. José García-Velasco and his collaborators at the University of Mexico School of Medicine had peered into the noses of one thousand randomly chosen subjects and found a VNO in almost every one of them.

The two American teams also outlined detailed anatomical studies, complete with photomicrographs to document their observations. Their studies showed that each vomeronasal pit leads to a closed tube 2 to 8 millimeters (from 0.1 to 0.3 inch) in length. These tubes are lined with cells reminiscent of those found in the vomeronasal organs of other species. Axons, which are the long processes that carry nerve impulses from sensory cells to the brain, are present among these cells. The results and photomicrographs left no doubt that normal human beings have a structure in their noses that looks very much like the functioning VNO of other mammals. The conclusion seems inescapable that human beings have an anatomical structure that could mediate a sixth sense. The universal presence of these pits and tubes had remained unrecognized through centuries of research on the human body.

If humans have a VNO, what does it do? Does it have any sensory function? If so, what does it detect? At the same Paris symposium, Luis Monti-Bloch and B. I. Grosser of the University of Utah School of Medicine addressed these fascinating questions. They had delivered puffs of air laden with several different chemicals to both olfactory and vomeronasal receptors of human volunteers and measured the response. When they applied a common odorant such as clove oil to olfactory receptors, the receptors responded vigorously. (This is no surprise. Clove oil smells strongly of cloves, and signals from these receptors carried the message, "Cloves!" to the brain.) In contrast, a puff of air carrying clove oil produced no response in the volunteers' vomeronasal receptors.

The responses of the two sets of receptors were reversed when Monti-Bloch and Grosser tested several other possible stimuli. This second set of samples caused strong activity in the VNO but had no effect on olfactory receptors. Several of these samples were effective in both men and women, but two were not. A sample designated ER-670 stimulated only women's vomeronasal receptors, while ER-830 stimulated only men's.

The second set of samples behaved just as one might expect for pheromones that are detected specifically in the human VNO. What were these fascinating samples, and where did they come from? Everyone wanted to know more about them. However, the results were as mysterious as they were sensational. No one learned anything more, because Monti-Bloch and Grosser merely designated the samples as "putative pheromones" and identified them by code numbers. They offered no further information beyond acknowledging that they had received the samples from a corporate source. Why had these investigators failed to reveal more about their sensational samples?

To answer that question, we must turn to a story that began more than thirty years earlier in Utah, long before the symposium in Paris. The story as we know it comes from its principal character, David L. Berliner. In the 1950s, Berliner was a medical scientist at the University of Utah School of Medicine, investigating the properties of chemical compounds obtained from human skin. Berliner had an endless supply of plaster casts removed from patients who had broken their legs. From the inside lining of the casts he could scrape all the dead skin he needed for his research.

As his work proceeded, Berliner noticed that certain skin extracts prepared from his scrapings had strange properties. When he left flasks of these extracts open in the laboratory, Berliner felt relaxed, confident, and secure about himself. Colleagues working beside him became more cooperative, more friendly with one another. When he removed the flasks from the laboratory, the scientists' mellow mood vanished; they reverted to their usual, grouchy selves. Berliner was intrigued by these skin extracts and their effects, but before he could explore them further, other matters diverted his attention. He had to terminate his skin investigations, so he sealed up the flasks of extracts and froze them.

Berliner returned to his frozen flasks only in 1989. In the intervening thirty years or more, he had left the University of Utah and had become a wealthy man. He now had his own laboratory, and it was here that he thawed his extracts and purified the strange chemical compounds from human skin. Berliner reasoned that his skin chemicals might be human pheromones. Chemical compounds that made people feel good about themselves would be a phenomenal discovery. Their practical applications would range from medicine to perfumery, and their commercial value could be enormous. For these reasons, Berliner was eager to publicize his compounds' effects but not to disclose their chemical identity. Until patents protected his discoveries, he would keep all chemical information secret.

To exploit his findings, Berliner organized a company he called the Erox Corporation. He also arranged for more research on his discoveries. This research included investigations into the biological properties of the skin compounds and studies on the human VNO. Finally, Berliner sought a forum where he could present his findings directly to the scientific community. The 1991 symposium in Paris was an ideal setting for this purpose, and the Erox Corporation contributed its support to this meeting.

At the Paris symposium, it was the Erox Corporation that Monti-Bloch and Grosser thanked for providing their VNO-active samples. From Berliner's story, it is now evident why only code numbers identified the Erox samples. They contained Berliner's skin compounds, and their chemical identity was Berliner's secret.

Over the next two years, Berliner's skin chemicals came increasingly to public attention. In 1993, the Erox Corporation launched two perfumes, one for women and one for men. According to Erox's advertising, each perfume incorporated a different human pheromone. Contrary to the hope and expectation of some, the pheromones were not sex attractants or aphrodisiacs. Instead, Erox said, their signals made the wearer feel better and more self-confident. The women's scent had a pheromone from women's skin that affected only women, and the men's scent had a corresponding pheromone from men. Presumably, these components were related to samples ER-670

170

and ER-830 that Monti-Bloch and Grosser found were vomeronasal-active only in one sex or the other.

The advertising also spelled out the benefits of these novel pheromone-containing perfumes. An improved self-image and increased assurance would naturally make the wearer more attractive to the opposite sex. Romance was sure to follow. Both *The Wall Street Journal* and *Vogue* magazine reported the entire story, from Berliner's early investigations in Utah to Erox's perfumes. Other publications, from *Fortune* magazine to *Women's Wear Daily*, ran portions of the story.

A further boost from the media came in the spring of 1994, when a television news team flew to Utah to report on Erox's research. In a laboratory setting, and with the television cameras whirring away, a white-coated scientist attached his test-apparatus to a reporter and administered three different samples, one after the other to her VNO. Only one of the three samples, the scientist explained, was a vomeronasal stimulant. After the third test, the reporter correctly identified the VNO stimulant without hesitation. Her report on Erox and the VNO was broadcast nationwide on a popular television news magazine. A few weeks later, the *MacNeil-Lehrer News Hour* presented an essay on pheromones. Prominently featured were Berliner, his work, and his perfumes.

As the Berliner story has unfolded in technical journals and the public press, the scientific community has reacted with cautious interest. Mood-altering human pheromones would be as exciting to interested scientists as to the public at large, but most scientists take a wait-and-see attitude toward claims that cannot be independently verified. There is no reason to disbelieve the technical papers that have appeared, but there is also no way to confirm Berliner's claims at present.

If the chemical nature of the now secret compounds were public information, other scientists could obtain the compounds and explore their vomeronasal activity. Even if the compounds are VNO stimuli, independent scientists will want to probe their behavioral effects. Tests with appropriate safeguards and controls are fairly elaborate, but they would assure that any mood changes observed indeed come from the compounds being examined. Until Berliner's claims are independently verified, most scientists will suspend judgment.

The scientific community's reservations seem unlikely to dampen public enthusiasm. Erox already has testimonials from elated users eager to disclose how the perfumes have transformed their personal lives. An independent company has announced its own pheromone-containing perfume. Meanwhile, Berliner has organized a second corporation to perfect other products based on his compounds from skin. Human pheromones could have a spectacular future.

CHAPTER

8

CHEMICAL MESSAGES
TO THE WIDER WORLD

*F*ROM BACTERIA to human beings, organisms communicate with their own kind by way of chemical messages, often sending signals that are vital for survival. What about messages from one species to another? What do unrelated organisms have to say to one another? Are these messages as significant as pheromones? We know less here than we do about pheromones, but one answer to these questions comes from an everyday scene on the Texas plains.

A prairie rattler strikes at a mouse; its jaws close, and the fangs sink in. For a moment, the snake holds the mouse, then releases it. The snake frees its prey after striking, but the terrified mouse will not escape its fate. While the mouse is running for its life, the snake's venom is already at work. In a few moments, the mouse sinks to the ground, its flight ended. Soon the rattler comes along, seizes the mouse in its jaws, and swallows it whole.

The hunting strategy of the prairie rattlesnake (*Crotalus viridis viridis*) is successful because the snake can track the fleeing mouse. As the mouse runs, it leaves an odor trail that the snake can easily follow. In effect, the trail is an inadvertent chemical signal from the mouse to the rattlesnake. Like other messages that identify a prey species to its predator, it is an interspecific signal that benefits only the receiver. The mouse passively "communicates" its odor to the prairie rattler but gains nothing by doing so: the consequence of the mouse's message is that it is soon devoured.

In contrast, there is nothing passive about the snake's performance. A prairie rattlesnake is superbly equipped to receive and act on the mouse's message, and it regularly depends on chemical messages from its prey to feed itself. Like other snakes, prairie rattlers have an efficient system for gathering these messages and a well-developed vomeronasal organ for processing them.

As the rattler follows the dying mouse's trail, it touches the tips of its forked tongue lightly to the ground and then retracts the tongue into its mouth. Odorants picked up from the ground pass through a pair of ducts in the roof of the snake's mouth and into the two fingers of its VNO. Here, receptor cells detect the odorants and send appropriate nerve impulses to the snake's brain. The rattler samples the ground with its tongue about forty times a minute as it glides along, and the odorants it senses guide it to the mouse. The critical information comes from the complex odor of the mouse's skin and fur. Although the fleeing mouse leaves visible amounts of urine on its trail, the snake ignores that odor.

MESSAGES BETWEEN HUNTERS AND THEIR VICTIMS

The rattlesnake and the mouse interact as predator and prey. Many encounters between species can be viewed broadly either as predator-prey interactions or as closely related activities, such as those involving a parasite and its host. We can ignore some technical differences and refer to this entire group of activities as hunter-victim interactions. These very common interactions often include an exchange of chemical messages. Such interactions provide much of what we know about interspecific chemical communication.

Various kinds of messages pass back and forth between a hunter and its potential victim. One that we have already talked about is primarily a chemical defense used by the victim against the predator. The skunk's repellant spray is an effective chemical warfare agent, but it is also a message that reminds would-be predators, "Keep away!" The parasitic beetles we discussed earlier also have a defense that is an interspecific message. They pass their lives comfortably in ant nests, protected from rude expulsion by a coating of the ants' own cuticular chemicals. For the beetles, these compounds are both a chemical defense and a message that declares to an in-quiring ant, "I am a member of this colony."

Another chemical warfare agent that also acts as a signal between spe-cies is a spray used by a slave-making ant (*Harpagoxenus sublaevis*) on its raids. These belligerent ants boldly invade the nest of a chosen ant-victim. They throw out many of the resident ants or simply kill them, and then carry the victim's brood back to their own nest and raise them as slaves. During their raids, the slave-making ants bite their victims and spray them with a noxious liquid. The ant-victims defend their nest, but the spray con-fuses and terrorizes them. Under its influence, they turn upon their own nestmates and begin to fight among themselves. The chemical spray has brought a message that promotes dissension and civil war.

Some hunters locate their victims by eavesdropping on the victims' own signals. The spined soldier bug's pheromone serves a secondary role of this sort. In addition to drawing other soldier bugs to a new food source, the pheromone broadcasts an unintended message to several other species, in-cluding some small yellow and black flies (*Euclytia flava*) that have a lifelong interest in soldier bugs. These flies lay their eggs on the bugs, and their maggots later feed on them. If a female fly detects the bugs' attractant, she joins the soldier bugs hurrying to its source. The pheromone that promises food and mates to the bugs means egg sites to the fly. She attaches her eggs to a soldier bug with a bit of glue, and when the maggots hatch, they bore into the bug and feed on it as they grow. When hunters read their victims' pheromones in this way, once private messages become interspecific signals. The parasite has a helpful means of finding its victims.

Many hunters that use chemical signals for locating or identifying their victims need additional information as well. There is a tiny parasitic wasp, *Telenomus heliothidis*, that deposits its eggs inside the eggs of tobacco bud-worm moths. The moth eggs carry a particular protein on their surface, and a female wasp looks for this protein to identify an appropriate site for her own eggs. However, the protein is not the only cue she requires. Entomolo-gists at Texas A&M University obtained the marker protein from female to-bacco budworm moths and painted it on glass beads of various sizes and shapes. They offered these painted beads, along with clean, unpainted beads, to a female wasp. The wasp examined and probed painted beads that were the same size and shape as the moth eggs, but she ignored both the clean beads and also painted beads of the improper size or shape. Apparently, the wasp looks for eggs that have the right size, the right shape, and the critical protein on their surface. She explores further only when she finds all three indicators together.

SOME OF the strangest hunter-victim interactions take place between predatory plants and their prey. Many kinds of herbivorous insects feed on plants, but a few plants have managed to turn the tables. Venus's flytraps and other carnivorous plants capture insects and feed on them. Among plant predators, one of the oddest that makes use of a chemical signal is shepherd's purse (*Capsella bursa-pastoris*), a common weed that grows around the world. The peculiarity here is that the seeds of shepherd's purse, not the mature plants, are carnivorous.

On immersion in water, the small shepherd's purse seeds release a chemical that draws the attention of minute insect larvae. The larvae swim up to the seeds and become glued headfirst to the seeds' sticky outer covering. As many as twenty larvae can become stuck to a single seed. The wet seeds also emit a toxin that kills the larvae and an enzyme that digests them, breaking down larval proteins to amino acids. The seeds then absorb the amino acids from the water and incorporate them into their own growth.

The predatory activities of shepherd's purse seeds came to light in laboratory experiments. The observations are extraordinary, but they do not tell us how the seeds behave in nature. That is, no one knows whether seeds lying in the damp earth really prey on insect larvae and other soil organisms. At the same time, shepherd's purse is one of the world's most widespread flowering plants, and it is tempting to suggest that its carnivorous seeds contribute to its success.

Another unlikely group of carnivores are fungi that attract and feed on microscopic forms of soil life. Many fungi consist of long microscopic filaments, or hyphae, that invade and digest wood and forest debris. They slowly convert the hardest wood to digestible sugars, returning this material to the pool of generally accessible foodstuffs. Without their constant work, much of the earth would fill up with dead trees and underbrush. Many of these fungi come to casual attention only when their so-called fruiting bodies appear above ground or on the surface of trees or rotting logs to bear

176

and release millions of spores. We know many of these fruiting bodies as mushrooms.

Although these fungi can feed on a never-ending supply of wood, their diet has the same disadvantage as a diet of pure carbohydrate has for humans: it provides very little nitrogen. Like all other living organisms, fungi use nitrogen for synthesizing their proteins and nucleic acids, and nitrogen comes largely from dietary protein. Many humans meet this nutritional requirement by including meat in their diet, and over one hundred wood-digesting fungi do the same. They supplement their diet with soil nematodes and other tiny animals rich in protein. The net fungus (*Arthrobotrys oligospora*) takes its common name from netlike traps attached to its hyphae. These structures release a chemical signal that lures soil nematodes from afar. When a nematode arrives, a powerful adhesive on the net-traps binds it to the fungus. Fungal hyphae invade the worm's body, carrying toxins that paralyze the worm and enzymes that dissolve and digest its substance, providing essential nitrogen to the fungus.

A related fungus (*Arthrobotrys anchonia*) traps its prey with a noose instead of an adhesive. It also uses a chemical lure, which emanates from little rings, each consisting of three cells, that are distributed along the hyphae. A nematode responding to the attractant may enter one of these rings. If its

8.1 The slim nematode caught in two contracting rings of the Arthrobotrys *fungus is being squeezed to death. Soon, threadlike hyphae will grow out from the fungus, invade the worm's body, and digest it.*

body should touch the ring's inner wall as it passes through, it triggers a deadly response. The three cells of the ring instantly swell and crush the hapless worm, which the fungus then digests.

Luring prey with chemical signals works well for these fungi, and many other hunters have adopted a similar strategy. Most simply, a hunter produces and disseminates a signal as bait, then waits for its victim to respond. The victim mistakes the signal for its own attractant pheromone, or perhaps for an inviting food odor. Innocently scurrying to it, the victim falls into the hunter's hands before discovering its error.

There are also biologically more elaborate ways of creating a chemical lure. One of these is a ploy developed by two predators that cooperatively broadcast an attractant and then devour the prey that respond. This joint venture is the work of a spider and some yeast species. With the spider's help, the yeast make an attractant that brings flies to the spider's web. The spider traps the flies and shares this food with the yeast.

Most spiders are solitary creatures, but a few live socially. One of these, a Mexican spider called *Mallos gregalis*, lives in huge colonies. As many as twenty thousand spiders of both sexes and all ages build giant webs that cover whole branches of trees.

The spiders' feeding habits are unusual. When most spiders feed, they suck an insect dry and discard the husk that was the animal's external skeleton. *M. gregalis*, however, does not completely consume a captured fly, nor does it discard the fly's carcass. Instead, it deliberately incorporates the fly-remains into its extensive web, spinning new silk to add a partially consumed fly here and there. Spiders are tidy creatures, and collecting dead flies is most unspidery behavior.

The fly-remains accumulated in the web have a surprising fate. Like other organisms, the living flies were host to a population of bacteria and yeast. Ordinarily, when an organism dies and its natural defenses are disabled, the resident bacteria multiply rapidly and begin to decompose the corpse. As the bacteria go about their work, the corpse takes on a rancid, ammoniacal odor. However, fly-remains that these spiders save and insert into their web do not turn rancid. Instead, these remains decompose with a sweetish, yeasty odor. This yeasty odor permeates the entire spider colony and is a powerful attractant for flies. The reason for the spiders' unspidery

178

habits is now apparent. By collecting fly remains, the spiders lure more flies to the colony.

However, the spiders are only partially responsible for their fly attractant. The odor comes from flourishing yeast that were resident in the living flies but are usually overwhelmed by the proliferating bacteria after the flies die. The spiders' contribution is to save the yeast by wiping out the bacteria. After the spiders feed on the flies, the fly-remains are still hospitable to the yeast. Presumably, the spiders inject an antibiotic when they kill the flies, neutralizing the bacteria and so permitting the yeast to thrive. Dead flies decomposed by yeast rather than bacteria do not turn rancid. As the yeast flourish and multiply, they generate the attractive odor that brings more flies.

Owing to cooperation with the yeast, colonies of these spiders are typically surrounded by swarms of flies. Many flies blunder into the webs and became entrapped, providing food for yeast and spiders. There seems to be an inexhaustible supply of flies, so this gratifying arrangement should continue indefinitely. Occasionally, however, the system falls out of balance. Should there be too many flies or too few spiders, the web may entrap more flies than the colony can handle. Some of the trapped flies die before the spiders feed on them. Without the protection of the spiders' antibiotic, the yeast are at the mercy of the bacteria. The bacteria proliferate unchecked in the fly corpses, destroying the yeast and decomposing the flies in their own way. The fly corpses turn rancid, and soon the entire spider nest smells bad. Without the yeasty attractant, flies no longer come in swarms, and the spider colony soon has too little to eat. As a consequence of trapping too many flies, there are now too few. The spiders have no choice but to abandon the polluted nest and rebuild their colony elsewhere.

STEALING SIGNALS FOR PEST CONTROL

In human terms, the most important chemical signals between hunters and victims are those involving agricultural pests and crop plants. In addition to signals between pests and the plants they feed upon, there are signals between beneficial insects and the crop pests that are their prey. Agricultural scientists are only beginning to decipher these interspecific signals, but their ultimate goal here is the one they have pursued for years: pest control that is practical, biorational, and environmentally safe. Because we are only beginning to learn about interspecific chemical signals, practical application of this research still lies in the future.

Several different chemical signals may pass between an insect pest and the crop plants it threatens. Most pests are specialists, and if they fail to find the right food, they cannot succeed. Both adults and larvae may feed only on particular plants, and not necessarily the same ones. Adults may lay their eggs only in certain sites, often chosen so that on hatching, larvae are near the food they require. To locate the right sites, specialist pests often rely on volatile chemical signals coming from the plants. Once they locate a plant, they may look for additional signals before feeding or depositing eggs. These insects sniff out their victims just as carefully as snakes or wolves do.

One economically significant pest of this sort is a beetle (*Diabrotica virgifera virgifera*) whose larvae attack the roots of corn plants. The larvae are known as western corn rootworms. Together with northern corn rootworms (*Diabrotica barberi*), they are the most serious corn pests in the north central United States and in Canada. Farmers control the larvae by dousing their fields with soil insecticides, and they control the adults through aerial spraying. The overall cost of these insecticides, plus the value of the crop lost to the pests, is about $1 billion a year. This figure makes no allowance either for the value of the countless beneficial insects incidentally exterminated, or for the burden of pesticide pollution of lakes and streams.

An annual billion-dollar expense suggests there is room for improvement in managing corn rootworms, and agricultural scientists have studied the beetles in some detail. One interesting finding concerns a chemical signal vital to the western corn rootworms' survival. The pests appear in the soil of cornfields during May and June, hatching from eggs that were laid late the preceding summer and that have then overwintered in the ground. For these newly hatched larvae to survive, they must find the roots of young corn plants. Unless they attach themselves to corn roots, they soon die. They move through the soil, guided toward their goal by a signal emanating from the roots, much as a male moth moves toward his female's attractant. For the corn rootworms, the attractant is simply carbon dioxide, which the roots normally release as they develop.

In laboratory experiments, corn rootworms move efficiently toward a source of carbon dioxide. (Bottled club soda provides entomologists a convenient source of the signal in the laboratory.) The larval receptor for carbon dioxide is apparently not very finely tuned to the size and shape of the carbon dioxide molecule. In a parallel set of experiments, the larvae also responded to other simple chemicals that roughly mimic the size and shape of carbon dioxide. Perhaps this mimicking effect can contribute to controlling these pests in the future.

While corn rootworms must seek out corn roots, there is another destructive corn pest that hatches directly onto an ear of corn, the food it desires. Female corn earworm moths follow volatile chemicals from corn silk to locate suitable sites for their eggs. Only after receiving this signal indicating an egg site, do they begin to synthesize and release their sex pheromone to attract a male moth. Thus, even before advertising for a mate, these moths choose a place to deposit the eggs that are to come.

Another pest that carefully positions her larvae to assure their future is the common European cabbage root fly (*Delia radicum*). The female fly searches out a cabbage plant (*Brassica oleracea capitata*) or other crucifer (broccoli or brussels sprouts, for example) to receive her eggs. Volatile chemicals probably direct her initial search, but after she lands on a leaf, she expects an additional signal. The fly's feet carry sensitive receptors for sulfur-containing compounds associated with crucifer odor and flavor. Once the fly's feet pick up these compounds, she knows this is an appropriate plant for her larvae. Now she is ready to deposit her eggs.

THE WAR ON WITCHWEED

As these examples show, interspecific chemical signals are beginning to find a place in the warfare on agricultural pests. Perhaps their greatest practical promise is in the continuing war against the tropical parasitic plant called witchweed (*Striga asiatica* and related species). Although witchweed is virtually unknown in the developed world, it is a constant companion of subsistence farmers all across the vast semi-arid tropics of Africa and India.

This destructive pest is a robust green plant with handsome pink flowers. A field of witchweed is quite lovely from a distance, but closer inspection reveals the ugly truth. The field of pink flowers was once filled with ripening grain, grain that is now stunted, bleached, and wilted. The flowers have essentially destroyed the crop, for the attractive plant cannot live independently but only survives attached to the roots of other plants. Through its attachment, the witchweed plant extracts from its host water, minerals, and sustenance for its own growth. While diverting host resources to its own needs, witchweed produces toxins that inhibit the host's development.

Witchweed attacks widely cultivated cereals and legumes such as corn, sorghum (*Sorghum bicolor*), pearl millet (*Pennisetum glaucum*), rice (*Oryza sativa*), and cowpea (*Vigna unguiculata*). Moreover, the list of victims grows continually, because the parasite adapts itself to new host species with frightening ease. These crops attacked by witchweed are the main source of energy and protein for hundreds of millions of people on two continents.

This is a phenomenally successful parasite, and it is responsible for catastrophic losses. In Africa, the overall average loss of the cereal grain crop to witchweed is about 40 percent. Heavily infested fields in Ethiopia and Sudan commonly lose from 65 to 100 percent. At this level of destruction, witchweed may be the greatest biological constraint on food production in Africa, perhaps causing more crop damage than plant diseases, birds, or insects.

Unfortunately, witchweed thrives best on plants already under stress from lack of moisture and nutrients. In tropical Asia and Africa, such conditions are not rare. They are commonplace in regions where the soil is thin and the climate dry. Infertile land cultivated by subsistence farmers tends to become heavily infested with witchweed. Farmers abandon such land as unproductive and migrate to new areas, where cultivation invites new infestations.

Changing ways of life have also worsened the problem in some regions. The parasite flourishes in fields that are cultivated steadily, season after season. Traditional crop rotation, long fallow periods, and slash-and-burn practices all once helped to keep witchweed in check. Nowadays, in many areas farmers are more permanently settled, and earlier lifestyles are disappearing. Where fields are now in continuous use, witchweed has emerged as a severe problem. The available evidence suggests that the problem is worsening. In countries where starvation is unexceptional, the war on witchweed can be a war for human survival.

Several characteristics of witchweed's peculiar parasitic existence contribute to its success. Each plant produces at least forty thousand seeds, and the number may reach five hundred thousand under favorable circumstances. The seeds are very small, about half the size of a grain of salt, and they remain viable in the ground for as long as twenty years. The seed from one witchweed plant can be enough to ruin an entire field.

Not only are the seeds long-lived, but they also have an unusually high probability of developing into a new plant. The seeds cannot germinate during the season they are produced but remain dormant in the soil until the following year. Then they await a period of rain that keeps the ground moist for a week or longer. This moisture brings the seeds out of dormancy, so that they are able to germinate.

Because witchweed is a parasite, it must attach itself to a host root within a few days of germinating, or it dies. Finding a host root is imperative, and witchweed leaves little to chance in doing so. The seeds actually germinate only on receiving a specific chemical signal from a potential host root. Like many other signals from victim to hunter, the host root's communication is unintentional. The witchweed seed merely responds to chemicals that are released by host roots during their normal growth, just as corn rootworms do. As a result, the only seeds to germinate are those that happen to be near a root of a potential host. Brief exposure to the germination signal is sufficient, and within twelve hours the seed begins to sprout. Seeds that germinate have a high probability of parasitizing the nearby host plant.

Seeds that could have germinated but receive no host stimulus simply return to dormancy and retain their viability for the future. Selective germination enhances each seed's likelihood of success and also builds a witchweed seed bank in the soil. A witchweed-infested field conceals millions of tiny timebombs ready to explode into new parasites whenever conditions are right.

After germinating, a witchweed seed extends a small rootlet toward the nearby host root. At its tip, the witchweed rootlet develops an organ known as a haustorium, which forms the physical attachment to the host root. More than one witchweed rootlet can invade a single host root, and of course the more parasites a host plant supports, the greater the damage it suffers. Development of the haustorium depends on a second chemical signal from the host root.

These properties of witchweed and its reliance on chemical signals make the plant a formidable foe. Eradicating witchweed may not be feasible, but bringing it under control is a sensible goal. A very promising effort in this direction is underway at Purdue University. Plant scientists,

183

geneticists, and biochemists there are combining basic research with practical measures designed to allieviate witchweed infestation in Kenya. In their research, these scientists are probing the basic interactions of witchweed with its host. A major thrust of their work is to decode the chemical signals that these interactions require. The Purdue investigators are then applying what they learn to develop new varieties of crop plants that are resistant to witchweed.

In an important step forward, the witchweed scientists have identified the stimulus from host roots that triggers germination of witchweed seeds. The roots of many plants exude several relatively complex compounds that are potent signals for witchweed. Like other chemical messengers, many of these compounds are effective in minute amounts. Working with sorghum, the scientists have also learned that the amount of chemical signal produced by various strains of a single host species can differ as much as a billionfold. Plant geneticists have used this information to create a new stable variety of sorghum that releases virtually no germination signal. Witchweed seeds lying in the soil near the roots of this new sorghum strain do not germinate, because they never receive a triggering signal. In field tests, the new Purdue sorghum shows good resistance to witchweed.

The Purdue scientists are also working on plants that stimulate "suicidal germination" of witchweed seeds. These plants release normal levels of a germination stimulus, but abnormally low levels of the haustorium signal. Witchweed seeds near the roots of these plants germinate normally and begin their growth toward the host root. However, the signal to form a haustorium never comes. Without this second signal, no haustorium develops, and the witchweed rootlet is doomed to an early death. It reaches its potential host but cannot form the attachment that is vital to its survival.

Another way to induce suicidal germination is to apply some external chemical that causes germination in the absence of any host plant. Witchweed is not very particular about the germination signal but germinates in response to a number of chemical compounds. Purdue scientists have identified several herbicides that stimulate germination and that are already on the market for other purposes. These compounds may offer a practical way to clean up fields that have been abandoned because of witchweed. However it is achieved, suicidal germination actively removes witchweed seeds from an infested field and so returns the field to cultivation.

In connection with the practical aspect of their program, some of the Purdue group visited Kenya a few years ago to learn about witchweed in the field. They hoped to talk directly with farmers who raise sorghum and fight the parasite every time they plant a crop. They chose a farming village, and the village men received them, ready to talk about witchweed. The conversation quickly revealed that the men acted only as overseers in the sorghum fields. The village women actually farmed the land and had the practical experience that interested the scientists. During this conversation between the men and the scientists, the village women had remained apart, many of them arriving at the meeting directly from their work in the fields. The village men were reluctant for the scientists to talk directly with the women, but one woman scientist in the group was finally able to speak with the female farmers. Unfortunately, the villagers lost interest in the meeting when they learned that the scientists had no solution to the witchweed problem.

If the scientists had nothing to offer on their first visit, they did better when they returned to Kenya in June 1994. On this visit, they brought the farmers the new witchweed-resistant sorghum they had developed at Purdue. This is the sorghum that performed well in field tests, and now the critical practical question is how well it performs over several growing seasons in the hands of local farmers.

Meanwhile, back in their laboratories, the witchweed investigators continue to examine the parasite's life cycle and to search for other chemical messages that pass between parasite and host. They have already learned that the witchweed plant's continued development after attachment to the host root depends on further chemical signals from the host. These later signals may offer new points of intervention and new opportunities to disrupt the parasite's growth and development. The plant scientists would like to create a crop plant that resists witchweed at each stage of the parasite's assault on its host. With multiple lines of defense, this ideal plant should survive and flourish unscathed by witchweed.

SIGNALS FOR SCHISTOSOMES

In many parts of Africa where witchweed contaminates the land, another sort of parasite strikes the people themselves. Wherever people enter lakes and streams contaminated with human waste, they may contract schistosomiasis. This widespread and debilitating disease has been endemic to

Africa at least since the Middle Kingdom of ancient Egypt, four thousand years ago. Schistosomiasis makes its home not only in Africa, but also in other tropical and subtropical regions of the earth. The Near East, China, the Philippines, parts of South America, and several Caribbean islands are all affected. Two hundred million people in at least seventy-five countries suffer from the disease, and 500 or 600 million more are at risk. After malaria, schistosomiasis is the world's most common lethal parasitic disease. It kills more than eight hundred thousand people each year.

This disease rages on despite the fact that drugs that can cure schistosomiasis have been available for several years. Not only is it is difficult to get these drugs to the people that need them, but those that are treated are also likely to be reinfected later. The drugs can cure the disease but cannot prevent reinfection. Like onchocerciasis, schistosomiasis finds its victims largely among impoverished rural people living under unsanitary conditions in countries where public health facilities are scarce. For these reasons, scientists continue to study schistosomiasis, testing new, more practical ways to control it. Their studies furnish a detailed picture of schistosomiasis and the parasites that cause it, including several chemical signals they use. Before talking about these signals and possible treatments for schistosomiasis, however, we must explore the truly bizarre life of these parasites.

The organisms that cause schistosomiasis are distant relatives of tapeworms and are known as blood flukes or schistosomes. While several species of these worms parasitize humans in slightly different ways in various parts of the world, we can discuss one species, *Schistosoma mansoni*, as representative of the entire set. (Additional species of schistosomes parasitize other mammals and also birds.)

Like many other invertebrate parasites, schistosomes lead complicated lives that proceed through a series of diverse forms. They spend part of their existence in a human host and part in one of several kinds of freshwater snails (*Biomphalaria* species). Male and female adult schistosomes live and lay eggs in humans, and an intermediate immature stage lives and multiplies asexually in snails. Schistosomes move back and forth between these two hosts in a life cycle that is filled with improbable events.

To follow this cycle, we shall start with a pair of mated adult worms. These two worms live together in the interior of a blood vessel that serves the liver of an infected human, feeding directly on their host's blood. As you might expect of animals that pass their lives in a blood vessel, schistosomes are worm-shaped, 1 to 3 centimeters (0.4 to 1.2 inches) long, but no more than 1 millimeter (0.04 inch) wide, about the size of a small pin or needle. The male is short and massive, and he has a sucker by which he attaches himself to the wall of the blood vessel. The longer, very thin female has paired with the male for life, and she resides permanently in an extensive groove that runs the length of the male's body. An average pair of worms lives several years in this intimate relationship, continuously producing and

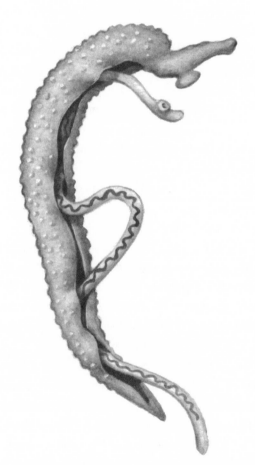

8.2 *The long, thin female schistosome of this mated pair lives nestled in the male's sex canal.* Schistosoma mansoni *is common throughout tropical Africa and spread to the New World tropics during the slave trade.*

fertilizing eggs. The female lays perhaps three hundred eggs a day, about one every five minutes, day and night. •

To continue the parasite's life cycle, an egg that is released in a blood vessel must be excreted in the host's feces. An egg must first make its way from the bloodstream into the host's intestine. There is no simple path for this migration. Somehow, the egg must get out of a blood vessel and work its way through connective tissue until it reaches the intestine. To facilitate the journey, each egg has a sharp spine on its surface. This spine may tear open a small vessel as the egg is swept along in the blood, permitting the egg to move out of the circulatory system and begin its journey to the intestine.

Needless to say, only a small fraction of the eggs succeed in making their way through the body's tissues to the intestine. Many never leave the circulation but are carried by the blood to the liver, lungs, and other sites elsewhere in the host's body. Eggs that remain in the host are irrelevant to schistosome reproduction, but they are the cause of the symptoms and pathological effects of the disease. Surrounded and encapsulated by the host's defensive cells, these eggs develop into cysts that eventually bring about enlargement of the liver and spleen, and degenerative changes in the circulatory system.

Within an egg, as it makes its journey through the host's body, the next form in the parasite's life cycle is undergoing its development. This next form, called a miracidium, comes into existence only if the egg leaves the host's body and passes into fresh water. As soon as a schistosome egg leaves the salt-containing fluids of the host's body and enters fresh water, the egg ruptures. This permits the fully developed miracidium, a small creature about half the size of a grain of salt, to swim free. If the egg fails to reach fresh water, it and the miracidium within soon die. Only where human feces pass into lakes and streams can schistosomiasis flourish.

The miracidium that swims free will survive only if the fresh water into which it was born is home to an appropriate kind of snail, and only if the miracidium finds one of these snails within the next ten or twelve hours. These are unlikely events, but the miracidium's only objective in life is to penetrate a suitable snail, and so connect the schistosome's existence in the human host with that in the snail. How this larva finds its snail host is not completely understood, but chemical signals have a role in this search that we shall discuss later.

Most miracidia doubtless fall prey to predators or otherwise perish without locating a snail. A few succeed. These miracidia enter snails and begin to proliferate at a fantastic rate. Their asexual reproduction in the snail leads through stages of development to the next form in the schistosome's life cycle. This form, called a cercaria, is released from the snail back into the water. Penetration by a miracidium probably shortens the infected snail's life, perhaps to thirty or forty weeks. During its remaining life, the snail becomes a factory for generating cercariae. As the schistosome's reproductive process goes forward, the snail releases about fifteen hundred cercariae a day. Hundreds of thousands of cercariae can result from infection of the snail by a single miracidium.

Like miracidia, cercariae are free-swimmming larvae. A cercaria has a body about 100 micrometers (0.004 inch) in length and a forked tail twice as long, so that overall it is roughly twice the size of a miracidium. The cercariae's goal is also to find a host, but the host they seek is a human. They are the bridge from the snail back to the human host. If cercariae do not meet some other earlier fate, they have about twenty-four hours to find and penetrate a human host in the water before they die. Chemical signals are significant in this search, and later we shall return to consider them.

Upon finding a human host, a cercaria penetrates the host's skin, often slipping down the side of a hair shaft. The body of the cercaria enters the host, but the tail is cast off to die. Life in fresh water and life in a host's body make different physiological demands on an organism, and after penetration, the cercaria undergoes rapid changes to adapt itself to its new environment.

The transformed cercaria then begins migrating through the host's body and developing into a worm. Within a week the immature organism reaches the host's lungs, and some days later it moves on to the liver. Here it begins to feed on blood and to grow into an adult schistosome. As male and female schistosomes mature, they form pairs. The pairs migrate to their final home in a blood vessel near the liver and begin laying their eggs. The cycle is complete.

To summarize briefly: in one form or another, the schistosome must pass through a series of major steps for its life cycle to succeed. It must first get out of the bloodstream and into the intestine and pass out of the human host into fresh water. Then it must find a particular kind of snail within a few hours, and finally it must leave the snail and find a human in the water within a day. Along the way, it must subvert internal defenses of both snail and human during its parasitic phases, as well as elude freshwater predators while moving between hosts during its free-living phases. Furthermore, the appropriate snails are particular about their habitat and do not live everywhere. They thrive only in still or slow moving fresh water that is neither too hot nor too cold. Finally, the human hosts do not live in the water at all but only enter it occasionally for limited periods of time.

With a lifestyle based on an incredible concatenation of highly improbable events, how do schistosomes survive? The answer is of course in the potentially vast number of offspring that a pair of adult worms generates. The various numbers cited above are inexact and are only conservative averages at best. Nevertheless, they can yield a very rough estimate of the potential number of offspring from one pair of worms. The numbers imply that in a three-year lifetime, one pair of adult schistosomes would engender 120 billion new adult worms if all their eggs, miracidia, cercariae, and other intermediate forms successfully performed their parts in the life cycle.

In contrast to these 120 billion potential offspring, only two new adult worms are necessary to replace the original pair when they die after three years of endless reproduction. To provide two new adults and maintain the present number of schistosomes, the entire series of events must succeed only twice in 120 billion possibilities over three years. It is this almost inconceivable number of chances to succeed that permits schistosomes not only to survive, but to prosper and spread to new snail and human hosts.

In view of the poor chances miracidia and cercariae have of finding appropriate hosts, it is understandable that these larvae take advantage of chemical signals to facilitate their search. Many investigators have observed that water in which snails have lived attracts miracidia efficiently. If one carefully adds a small volume of "snail water" without mixing to a receptacle of ordinary pond water containing miracidia, the larvae quickly congregate and remain in the treated water.

Snail water attracts miracidia, but it is not a very specific response. Miracidia respond to more than sixty chemical compounds that are commonly excreted by snails and other organisms. The snail-water response does little more than orient the larvae to living organisms. As a result, all sorts of creatures attract miracidia, but only a few of these are suitable hosts. Miracidia also react to localized chemicals released by the snail once they are in contact with a potential host. These signals must also be rather nonspecific, because miracidia frequently penetrate snails of the wrong species, where they die.

Cercariae seem to use both temperature and light as signals in locating their human hosts. Chemicals have no known role in this first step, but they are important once cercariae make initial contact. Then, at least three sequential signals influence the cercariae's behavior. We owe our knowledge of these signals and their messages to research during the 1980s by Wilfried Haas and his collaborators at the University of Erlangen in Germany. The first signal causes a cercaria to attach briefly to the host's skin. If the cercaria does not detect the second signal within one or two seconds, it drops off and abandons the potential host. However, if the cercaria does encounter the second signal, it begins to creep along, exploring the skin surface. Neither of these two chemical signals is absolutely necessary, because cercariae also attach themselves and explore the skin in response to its temperature. In contrast, the third chemical signal is an essential part of the life cycle. Only on detecting this third signal does the cercaria penetrate the host's skin.

Haas and his colleagues have deduced the chemical nature of all three of these signals. The signal that encourages brief attachment is the amino acid known as arginine. Like other common amino acids, arginine is present in trace amounts on human skin. This indicator confirms to the cercaria that it has encountered a potential host.

Arginine is a sensible attachment signal, because it is normally present on human skin. However, there may be a more interesting reason for its role here. Arginine may be both an interspecific signal and also a pheromone. That is, the arginine signal may come both from the human host and also directly from other cercariae. This interesting possibility exists for the following reason.

A cercaria's tail is exceptionally rich in arginine, and as a cercaria penetrates its host's skin, the discarded tail marks the site of penetration with a telltale spot of arginine. This accidental bit of arginine may have evolved into a message, so that successful penetration by one cercaria encourages the attention of other, later arriving cercariae. The newcomers should pause to consider this site as a possible host, because another cercaria has already penetrated here.

8.3 A cercaria plunges into a mouse's skin adjacent to a hair shaft. The cercaria's head has already disappeared beneath the skin surface, and the discarded tail will soon drop off.

The arginine signal is a single compound, but both the second and third signals recognized by cercariae are chemically more complex. Each of these signals is a mixture of compounds found among the many lipids (fats) present on the skin surface. While we need not discuss these lipids in chemical detail, we should note that the two signals depend on mixtures of quite different compounds. For the cercariae the signals are easily distinguishable. It is also important to note that organisms other than humans carry these lipids on their outer surface. Like the nonspecific chemical attractant for miracidia, the three signals that guide cercariae may lead them to explore and penetrate not only human beings but also other, inappropriate hosts.

The third signal is particularly interesting. This lipid complex is essential for penetration, and the message it bears sets into motion far-reaching, irreversible changes. On sensing this signal, cercariae commit themselves to a host. They burrow into the hosts's skin, shed their tails, and undergo fundamental biochemical changes. They transform themselves from free-living forms adapted to fresh water into parasites able to live in the host's body fluids. Hereafter, a cercaria can no longer swim, because it has lost its tail. It can no longer even survive in fresh water but is obliged to live as a parasite. Few chemical signals bring about such irreversible changes in any organism.

Cercariae respond to this third signal even when they detect it on an inappropriate surface or in the water. Under such unnatural conditions, the penetration signal becomes a call to suicide, as a simple experiment demonstrates. A small bit of skin oil (from a person's forehead, for example) is dabbed onto the inside bottom of a glass laboratory dish, and pond water containing cercariae is added to the dish. The larvae quickly collect at the spot of oil and try to penetrate the bottom of the dish. At the same time, they shed their tails, and biochemical alterations take place in their bodies. Of course, the cercariae cannot penetrate the bottom of the glass dish. They remain in the water but are now unfit to survive outside a host. In response to the penetration signal, they have prepared themselves for an unattainable future, and they die quickly. If this lipid signal is dispersed in the water, it triggers the same suicidal behavior even in the absence of a surface to penetrate.

The lipids that carry this third message are chemically simple, and cercariae are not very demanding about the nature of the signal. Consequently, several ordinary lipids commonly found in chemical laboratories can successfully substitute for the natural mixture. This has led Haas to suggest that such compounds might offer a practical way to control schistosomiasis. Waters that are home to schistosome-infected snails could be sprayed with inexpensive lipids to destroy cercariae and limit future infection of humans. The compounds that deliver the lethal message to cercariae are generally nontoxic to other organisms, and so they should not produce unwanted side effects. It is also unlikely that cercariae would develop resistance to these compounds, since the natural lipids are an indispensable signal in their life cycle.

Adult schistosomes in their human host have their chemical signals, too. One is a lipid released by male worms, particularly before they have paired with females. In laboratory experiments, this lipid attracts female worms. In the body of an infected human, the lipid is probably an attractant pheromone that facilitates pairing of the worms.

Finally, a chemical signal from the human host stimulates adult worms to lay eggs. Investgators at the University of California in San Francisco were surprised in 1992 when they identified this signal as a familiar protein known as tumor necrosis factor alpha (TNFα). TNFα is a normal component of the human defensive response provoked by schistosome eggs. It is one of the internal signals used by the host in deploying its defensive cells against the eggs. The surprise here is that one of the battle signals the host employs to fight schistosome eggs actually stimulates the worms to lay more eggs. Schistosomes have adapted themselves so perfectly to life as human parasites that they have turned their host's defensive system to their own use instead of succumbing to it.

This discovery suggests that drugs that suppress the body's synthesis of TNFα might be useful in treating schistosomiasis. If the body synthesized less TNFα, schistosomes would be less stimulated to lay eggs. Fewer eggs would reduce the pathological effects of infection. An objection to this approach to controlling schistosomiasis is that the optimal level of TNFα is difficult to assess, so that reducing TNFα could do more harm than good. This is because TNFα also protects the host against the very schistosome eggs that it stimulates the adult worms to lay. The proper level of TNFα that effectively limits egg production, while still protecting against those eggs that are produced, requires a delicate balance. This balance could be difficult to maintain in routinely treating humans suffering from schistosomiasis.

Neither the idea of limiting TNFα production nor any other treatment of schistosomiasis based on chemical signals has yet received extensive study, although medical and public health officials have discussed a variety of fresh approaches to fighting the disease. The idea of a vaccine has been popular for years, because vaccination could offer lifetime protection against reinfection by new cercariae. Despite much research and testing, no vaccine is yet available for large-scale use.

Quite a different approach is to stamp out schistosomiasis through reducing snail populations. The idea is not new, but a novel possibility appeared in the early 1990s. Epidemiologists noticed that the area around Lake Malawi in Africa was relatively free of schistosomiasis until 1992, when the disease struck the local population. They discovered that this outbreak followed a decline in the lake's population of a snail-eating fish, *Trenatocranus placadon*. This decline in the fish population resulted from overfishing by local villagers. As the number of fish in the lake decreased, the population of schistosome-infected snails rose, and schistosomiasis began to infect the villagers. In 1994, public health officials began to consider restocking Lake Malawi with the fish to bring the number of snails back down to a safe level. If they pursue this strategy, one of their first steps must be to talk with the villagers about restricting fishing once the fish population is replenished.

Schistosomiasis is complex, both medically and socially, and schistosomes lead a life complicated by both a sequence of forms and unusual events. This complexity may offer public health officials multiple points of attack to confront and control the disease. Until new approaches to schistosomiasis are forthcoming, however, control must rely on long-term measures that are essentially unaffordable for the poorest countries: education, construction of sanitary water facilities, and drug-delivery programs that reach the millions of people who are infected or at risk.

MESSAGES FOR COOPERATION

While hunters and their victims provide one rich source of chemical signals, cooperative arrangements between different species provide another. One such arrangement that interests agricultural scientists involves plant roots and soil fungi. The spores of certain fungi, called vesicular-arbuscular mycorrhizal (VAM) fungi, are in the soil everywhere. The ones that interest us here grow and reproduce only within living plant roots. (Their complicated name simply describes the physical appearance of the attachment between the fungus and the plant root.) Most of these fungi are not specialized but successfully invade the roots of many different species.

The resulting plant-fungus interaction is complicated and not completely understood, partly because it is difficult to study the fungi outside their plant hosts. It is clear, however, that here both partners benefit from the arrangement. The plant furnishes sustenance to the fungus, and in return, the fungus extends its hyphae out from the plant's roots to bring the plant soil nutrients more efficiently, effectively extending the root system.

Such associations are quite common, and around the world about 80 percent of all plants, whether growing naturally or in cultivation, live in association with VAM fungi. Most plant roots we pull from the ground are actually the mixed root systems of plants and their associated fungi.

These soil fungi make their plant hosts hardier, and plant scientists hope to exploit them to improve agricultural crops. With this in mind, Gene R. Safir and his colleagues at Michigan State University searched for chemical signals that might stimulate VAM fungi to invade white clover (*Trifolium repens*). They were able to identify two chemical compounds released by clover roots that attract the fungi.

The Michigan State group synthesized these attractants in the laboratory and then tested their effects on the plant. Clover grew and spread better in soil containing the compounds than in untreated soil, but only if the soil also contained fungal spores. If the investigators first sterilized the soil to kill all its spores, the added compounds did not affect the clover. When spores were present, the added compounds reinforced the natural signals emitted by clover roots and brought more fungi to invade the roots. This resulted in increased clover-fungus interaction and enhanced growth of the clover.

In 1992, Safir and his colleagues received United States patents protecting their discoveries. The hope is that their compounds will make other crops hardier as well, because clover is not the only crop plant to benefit from VAM fungi. Safir's compounds may also benefit crops as diverse as onions, rye, and soybeans.

Another cooperative interaction among species we have discussed is the pollination of flowers by insects, birds, and other animals. We noted the importance of color in attracting pollinators, and now we must note the parallel role of fragrance. Appealing scents are an additional way for plants to advertise for assistance. Lured by a fragrance that may resemble a food odor or even an attractant pheromone, animals come, take a bit of pollen or nectar, and pollinate a flower or two as they move about.

Floral fragrances attract at least some insects almost as effectively as mating pheromones do. Detailed information is scarce, so the behavior of oriental fruit flies (*Dacus dorsalis*) may not be typical. The male flies are very responsive to a component of several pleasant flower fragrances known as methyl eugenol. In experiments at the University of Illinois, the flies were nearly as sensitive to this compound as male silkworm moths are to bombykol. Male fruit flies detect and then fly to a source of methyl eugenol as much as half a mile away.

The flies are not merely attracted by methyl eugenol but display an insatiable appetite for the compound. Flies can be dislodged from a surface coated with methyl eugenol only by force. Given unlimited access to the compound, a fly will feed until it dies.

Many other insects find blossoms with a sweetish scent enticing, although they do not typically respond so avidly. Some other creatures prefer odors that humans find less pleasant, and to accommodate them, there are flowers whose scents are definitely not agreeable to us. Many flowers pollinated by bats, for example, have a strong musty aroma. Those that delight flies may have the fetid odor (and even the appearance) of decaying meat. In the redwood forests of Northern California, a tiny orchid, the heart-leaved twayblade (*Listera cordata*) is usually pollinated by fungus gnats. Its odor impresses even experienced botanists, who describe the smell as "truly repulsive," like "molluscs beginning to go bad."

CRUCIAL COOPERATION BETWEEN FIGS AND FIG WASPS

There is no reason, of course, that signals passing from plants to insects need to be either repulsive or pleasant to our noses. Carbon dioxide has an innocuous odor, but it is the absolutely essential signal in an extraordinary alliance that exists between fig trees and fig wasps. The wasps pollinate the trees' flowers, while the trees provide the setting for the wasps' entire life. This exchange of services is not optional, because neither fig nor wasp has an alternative way to meet its needs. Neither one can live without the other. The signal is simple, but the interaction it supports is amazingly complex.

There are more than nine hundred species of fig (species in the genus *Ficus*). While we shall call them fig "trees" here, there are not only trees of all sizes but shrubs and climbers as well. Nearly every one of them has its own species of wasp (all in the fig wasp family Agaonidae) that pollinates it and no other fig species. The fig wasps associated with fig trees are tiny creatures, none more than a few millimeters in length. Despite their small size, they have been recognized since classical times, and Aristotle discussed their role in propagating figs. Nonetheless, intimate details of this association came to light only in the 1970s. The long delay may owe something to the technical difficulties of observing creatures only a few millimeters long as they pursue their daily lives in the dark interior of a fig.

Many biologists have been fascinated by figs and fig wasps, but the details that interest us emerged largely from studies by Jacob Galil and his coworkers at the University of Tel Aviv in Israel. Galil's work ranged over several species of fig tree and their accompanying wasps. For many pairs of tree and insect, the significant details of the association are similar, and our account comes from studies on a number of species. (The details are considerably different for the edible Mediterranean fig [*Ficus carica*], which has developed unique habits during millennia of cultivation. In particular, the figs that we eat do not interest fig wasps.)

What we wish to explore are the activities of fig wasps inside figs, which are the fruit of fig trees. Before doing so, we must note that a fig is unlike other fruits. Fig trees are an ancient group of plants, and their fruit and flowers develop in a peculiar way. A fruit such as an apple or an orange or a holly berry is a body that develops after pollination and encloses the plant's seeds. A fig may or may not eventually contain seeds, but it arises long before pollination. Unlike ordinary fruits, a fig is initially a protective container for the plant's flowers, not its seeds.

This protective container is a cup-shaped structure whose lip turns and reaches across the top of the container to cover the opening. The turned lip closes the opening almost completely. It does not quite come together in the center, so that a small hole remains in the top of the fig. This hole is completely filled with overlapping scales that protect the interior of the fig. Inside the fig, hundreds of very small, closely packed flowers extend from the inner wall of the fig toward the center. The flowers and their stalks fill much of the interior, but there is an open space in the middle that extends up to the hole in the top. The flowers grow so densely, side by side, that they form a continuous mass. The open space in the middle is bordered by the tops of hundreds of minute flowers, both male and female. As in other plants that bear separately sexed flowers, the male flowers produce pollen, and the female flowers, after they have been pollinated, produce seeds.

Pollinating the fig flowers is the business of the female fig wasp. She comes to the fig tree not only to perform this service, but also to lay her eggs. A female fig wasp, laden with pollen and eggs, seeks out an appropriate fig for her attention. Guided by chemical signals, she passes over and ignores fig trees of other species that are pollinated by other kinds of fig wasp. Even if several kinds of fig trees are growing in one locale, she flies specifically to her favored species and selects a fig at the proper stage of maturity.

After selecting her fig, the wasp must get inside it. The small hole in the top of the fig is her entrance, but the way is blocked by the overlapping scales. The wasp pushes her broad, flat head against the scales. She is perhaps only 1 or 2 millimeters long and must struggle to force the scales aside. They do not yield easily, and, despite her efforts, the wasp may fail to gain entry. It is not unusual for there to be several dead wasps outside a fig, or part way in, among the scales. Often, only one wasp succeeds in reaching the interior of a fig, although several may have tried. Even a successful wasp does not enter the fig unscathed. She loses her wings in the struggle to push her way in, and often her antennae break off as well. These losses do not deter her. She needs neither wings nor antennae for her work inside the fig, and there will be no return trip to the outside world.

The injured wasp makes her way down into the fig. The male flowers are immature and do not yet bear pollen, but the female flowers are fully developed. The wasp crawls around, moving over the tops of the flowers, and begins to lay her eggs. Behind her, the scales reclose, once again sealing the opening through which she entered.

About half of the female flowers that the wasp encounters have long stalks and about half have short ones, but from her position atop the flowers, she cannot tell the two apart. She inserts her ovipositor indiscriminately into female flowers, both long and short, boring down the stalk as far as she can into the depth of the flower and then depositing an egg. In the short flowers, her ovipositor reaches all the way to the ovary at the bottom. In these flowers, the wasp implants her egg in the ovary, where it begins to develop. In the longer flowers, the ovipositor does not reach the bottom. Here, the wasp's egg is deposited in the stalk above the ovary and does not live.

While she is laying her eggs, the wasp is also pollinating the fig flowers using pollen she brought from the outside world. She arrived with two containers, called pollen pockets, filled with pollen, and now she distributes a few grains of pollen to each female flower as she moves among them. Each of her pollen pockets is about 0.2 millimeter long and fitted with a moveable lid. One pocket is on each side of the wasp's body, recessed between her first and second pairs of legs. She withdraws grains of pollen with her forelegs and dusts them on the flowers. The covers on the pockets prevent pollen from being lost as the wasp forces her way into the fig.

These grains of pollen fertilize the long flowers, whose ovaries are too distant to have received a wasp egg. Fig seeds begin to develop in these ovaries. In contrast, each of the ovaries of the short flowers that carries a wasp egg becomes a nursery for a developing wasp. When the fig wasp deposited her egg, she also left a chemical that causes the ovary to enlarge and develop new tissue. This enlarged ovary, which is called a gall, serves as both food and protective covering for the wasp larva that soon hatches from the egg.

Meanwhile, the female wasp has finished depositing her several hundred eggs, and she has dusted her supply of pollen over the female flowers. In return for expertly pollinating the long flowers, she has appropriated the short flowers for her brood. In effect, the fig tree concedes half its flowers to the wasp in "payment" for her services. Having disposed of her eggs and pollen, the wasp has fulfilled her tasks. She dies presently in the interior of the fig.

The stage is now set for the next set of events. Over the weeks that follow, the entire metamorphosis of each new wasp takes place inside its gall deep within the fig, from egg through larva and pupa to adult. Finally, the enclosed male and female wasps reach adulthood. The male wasps emerge from their galls first, biting their way out through the wall of tissue. The males are much smaller than the females, perhaps less than a millimeter long. They have no wings and in general look nothing like ordinary wasps. They are quite active, however. Each male crawls around excitedly, looking for a gall that contains a female wasp, which he probably identifies by a chemical marker. When he finds a female's gall, he gnaws a hole in the wall. He inserts the apex of his abdomen through the hole and impregnates the female in her gall. This union takes only a few minutes, and the male then withdraws and wanders off to look for more females. Later, the female wasp will enlarge the hole made by the male and emerge from her gall.

After seeking out and impregnating all the females they can find, the male wasps gather together and collectively burrow through the wall of the ripened fig to the outside world. When they complete their tunnel, they either crawl out of the fig and fall to the ground, or else they withdraw into the interior of the fig. In either case, the male wasps die. Their short life has consisted of biting, digging, and mating within the fig into which they were born.

While mating has occupied the new generation of wasps, the fig has been ripening. Its male flowers are now mature and covered with pollen. The agile, delicate female wasps now emerge from their galls and move to

200

the male flowers. Using combs on their legs, they scrape pollen from the flowers and fill their pollen pockets.

Once a female has loaded herself with pollen, she is ready to leave the fig. Attracted by light from the outer world, she makes her way through the tunnel burrowed by the males and out of the fig. She must now find a young fig that is ready to receive her eggs and pollen, and so complete the cycle. She will live only a few days and probably does not feed before fighting her way into the fig she chooses.

The timing of these strange events is the key to success for both fig and wasp. Male wasps must emerge before the females do, but not before the females are ready to mate. Female wasps must not leave their galls before the male flowers mature, or there will be no pollen for them to carry away. Further, the fruit must be soft and ripe enough for the male wasps to burrow through the wall, or else the females will be fatally trapped inside the fig. At the same time, the fruit must not ripen too fast, or it might fall to the ground before the larvae complete their metamorphosis. The wasps could not mature and would die without leaving their galls.

Obviously, the wasp and the fig must coordinate their growth and development by some mechanism that keeps the male wasps, the female wasps, and the ripening fig all on schedule. They solve this problem very economically by means of a single chemical signal of variable intensity. As the fig grows and ripens, the concentration of carbon dioxide inside the fig increases as a natural consequence of its development. The gas released during ripening is trapped within the fig, because the opening in the top reclosed after the female wasp entered.

Carbon dioxide is an effective signal because it affects male and female fig wasps in different ways. Male wasps are active at high concentrations of carbon dioxide and inactive at low concentrations, while females react in the opposite fashion. These dissimilar responses control the unfolding of events.

As the fig develops and the concentration of carbon dioxide rises, the wasps are undergoing metamorphosis in their galls. By the time the wasps reach maturity, the carbon dioxide level is high enough to keep the females inactive. The males, however, are vigorous in this atmosphere. They emerge from their galls fully active and proceed to mate with the sluggish females.

201

About two days before the male flowers mature, carbon dioxide reaches its maximum concentration, perhaps as great as 10 percent. The mated females remain inactive in their galls, while the males are busy tunnelling through the side of the fig to the outer world. As soon as the tunnel is complete, the fig is no longer sealed. Carbon dioxide begins to escape to the outside, and its concentration within the fig slowly drops.

When the concentration has fallen below 2 percent, the females rouse themselves and gnaw their way out of their galls. If everything has remained on schedule, when the females emerge, the male wasps are dead and the male flowers have matured. Pollen is available for the active females to harvest. The lower carbon dioxide concentration also facilitates final ripening of the fig. Over the next two or three days, as the female wasps depart, the fig turns soft and dark, and its seeds reach maturity.

Such complex cooperation requires good communication and numerous adaptations by both fig trees and fig wasps to each other. Among their more remarkable adaptations are those that enable carbon dioxide to coordinate the maturation of insect and fruit.

DECEPTIVE COOPERATION

Fig trees and fig wasps need each other to propagate their species. There is no question that each benefits from their alliance. Not all arrangements between two species are so fairly balanced as this. Occasionally, an exchange that appears to be straightforward and mutually advantageous really contains an element of deception.

Such deceptive cooperation is a way of life for a certain stinkhorn fungus, *Mutinus caninus*. Like many fungi, stinkhorns live underground and occasionally send spore-producing fruiting bodies above the surface. In this particular stinkhorn, these fruiting bodies are slender structures 7 to 10 centimeters (3 to 4 inches) long, topped with a slimy greenish-brown mass known as the gleba, which contains the spores. The gleba has a foul, fetid odor that gives stinkhorns their name. As in the twayblade orchid with the repulsive scent, this odor is a chemical signal. It efficiently attracts a variety of flies that ordinarily live or lay their eggs on carrion or feces.

The flies that visit the stinkhorn find the gleba delectable. It is one of their favorite foods, and they consume it enthusiastically, often cleaning it completely from the top of the stinkhorn. As they feed on the gleba, the flies ingest the microscopic spores it contains. The spores pass through the flies' digestive tract unchanged and so are disseminated in the flies' feces, or fly specks. Biologists have estimated that a single fly speck can carry more than 22 million fungal spores. Stinkhorns generally depend on these flies as the primary distributors of their spores.

The transaction between stinkhorns and flies looks like a fair exchange of services. The fungus advertises with a tempting odor and offers the flies a popular food. The flies take the food and spread the fungal spores. The flies seem satisfied with this arrangement, but in fact they are making a poor bargain. Entomologists at the University of Massachusetts offered female black blow flies (*Phormia regina*) a diet of stinkhorn gleba. The flies fed happily, but when they ate only gleba, the eggs they subsequently laid failed to develop properly. On a diet of liver, the same flies produced normal eggs.

The reason for this difference is not hard to find. The flies put a large amount of nitrogen into their eggs, and this nitrogen comes from protein in their diet. Black blow flies usually feed on a high-protein diet of fresh or decaying meat. The stinkhorn's gleba, however, contains very little protein. Feeding only on it, the flies get too little protein to lay normal eggs.

The flies are ignorant of nutritional requirements and protein deficiencies. They prize the gleba as a food owing to its odor or taste, not its protein content, but in doing so, they endanger their own reproduction. As long as the blow flies are satisfied and continue to distribute the fungal spores, the stinkhorn has no incentive to improve its offer. Protein is metabolically expensive to synthesize, and the stinkhorn saves energy by producing tasty but nutritionally deficient gleba.

In another association similar to the one between stinkhorns and flies, a chemical signal that works as a laxative is an important factor. Deception is probably a key element in this exchange as well, although that point is less certain in this case. The association is one that flowering plants and herbivorous animals frequently establish. The plant offers the animals its fruit, which are morsels of food, perhaps pleasantly fragrant and brightly colored. Animals accept the offer and feed on the fruit. In doing so, they swallow the plant's seeds, which they then disseminate in their feces.

Many plants have their seeds dispersed in this way, relying on the color, flavor, and scent of their fruit as chemical signals to entice herbivores. A cooperative and mobile animal can spread a plant's seeds efficiently, but the arrangement presents a troublesome problem. Before they can be dispersed, the seeds must pass through the animal's digestive system, where the digestive process may destroy them. Seeds that are disseminated in feces typically resist digestion, and, in fact, some must pass through an animal's digestive tract before they can germinate. Other seeds, however, are more sensitive. The longer they remain in the animal, the less likely they are to be viable when finally released. Even though these seeds resist complete destruction by digestive enzymes, too long exposure to these conditions can deactivate them.

With this in mind, biologists proposed some time ago that plants might well be incorporating a laxative in their fruit to speed the seeds' passage through an animal. In mid-1994, K. Greg Murray and his wife, Kathy Winnett-Murray, two ecologists at Hope College in Michigan, supported this proposal with some clever observations. The ecologists and their students worked with a shrub and a small bird that live in the rainy mountains of Costa Rica. The shrub is called *Witheringia solanacea*, and its fruit is a red, cherry-sized berry. The bird is the black-faced solitaire (*Myadestes melanops*), a smallish slate-gray member of the thrush family and the primary distributor of the shrub's seeds in the area the Murrays studied.

To investigate how the shrub's fruit affects the bird's digestion, the Murrays created an artificial fruit to feed the birds. They replaced the natural fruit with a jelly that contained the shrub's seeds, colored the jelly red, and molded it into small balls to resemble the fruit. Then they soaked a portion of the artificial berries in an extract of the natural fruit. In this way the ecologists created two batches of artificial fruit that differed in only one respect. One batch contained juice and soluble chemicals from the natural fruit, and one did not, while both contained the real seeds.

The Murrays then offered the artificial berries to solitaires. One group of birds received berries soaked in the natural extract, and another received untreated berries. Both groups ate the jelly and its seeds just as they would the real fruit. Solitaires that received artificial berries with fruit extract excreted half the seeds they ingested in about fifteen minutes, the same as when they fed on the real fruit. Solitaires that received artificial berries without fruit extract retained the seeds longer. They dropped half the seeds in about twenty-five minutes. This ten-minute difference indicates the fruit extract contained a laxative that caused the birds to discharge the seeds more quickly.

A difference of ten minutes in the bird's gut may seem too short a time to matter. However, it turns out to be critical for the seeds. After fifteen minutes in the solitaire's gut, 70 percent of the seeds were still viable, but after twenty-five minutes, only 20 percent were viable. The quicker passage through the bird preserved the viability of 50 percent of the seeds the birds ate.

The shorter residence time saves seeds that may later germinate and lead to new plants, so the laxative undoubtedly helps propagate the shrub. However, what about the laxative's effect on the birds? For them, its effect may well be negative, because the laxative may reduce the food value the birds extract from the fruit. If the fruit remained longer in the bird's digestive system, the bird might benefit more from it. Here is where the possibility of deception arises, because what the shrub offers the bird may be worth less than it seems. It is likely that the shrub deceives the solitaire, but that will be certain only with better understanding of the bird's feeding habits.

A laxative-containing fruit comes as no real surprise. Humans have been eating various fruits for their purging effects for a long time. The fruits we eat probably contain laxatives as well, and they may be present quite specifically for the plants' benefit. No one knows yet what chemical compounds in the shrub's fruit are responsible for the laxative effect, but answering this question is an important next step. In a more practical direction, there is a real possibility that this research could lead to a new natural laxative for human use.

9

MAKING USE OF
NATURE'S MOLECULES

WE SET OUT to explore nature's storehouse of chemicals, to discover what they do and what importance they hold for us. We soon found that the store of natural compounds includes contributions from organisms of all sorts. Chemicals figure in the lives of organisms from fungi to flowering plants, from bacteria to mammals. With chemicals, plants fight herbivores, mites spread alarms, and fireflies generate light, while moths lure their mates, and snakes locate their prey. Other organisms do a thousand other things. One particular chemical may serve several species in quite different ways. Another compound may be the creation of a single species, never appearing elsewhere in nature, as far as we know. One chemical may be absolutely vital to some creature's existence, while another may play a helpful but secondary role.

These special chemicals are everywhere in nature, and with them, organisms reach out to one another in friendly or hostile fashion. In a very real sense, these molecules unite the living world and make possible life as we know it. Whatever their functions, they fulfill them efficiently. Deployed for countless generations and modified through the fine-tuning of evolution, each of these chemicals is specifically adapted by nature to some creature's need.

We also found that human use of nature's chemicals began with what we called natural preparations. Probably starting with drugs derived from plants, people everywhere began concocting beneficial preparations from nature thousands of years ago. They worked by trial and error, and often credited their success to divine intervention. From our present perspective, they were making effective use of natural chemicals to change their lives. Natural preparations have continued to transform human existence down to the present, and now it is appropriate for us to inquire about their role in the years to come. We end our explorations with a brief look at the future.

Natural preparations were commonplace long before the first glimmerings of scientific thought, but today the development of new preparations goes hand in hand with scientific research. The discovery and use of new natural chemicals is typically the fruit of careful investigation, while the funds that support these investigations increasingly depend on the prospect of commercializing the chemicals discovered. Scientists' investigations of natural compounds are thus linked ever more strongly to practical application. This relationship need not hinder imaginative research, as long as those who control the funds recognize that scientific findings with practical promise turn up unpredictably. As we have seen, ecological studies on fruit-eating birds in Costa Rica may eventually give us a new natural laxative, while the construction practices of marine bristleworms may point the way to a marketable underwater adhesive.

Discovering other natural chemicals with practical applications should be a growing enterprise for the foreseeable future. Scientists have searched for chemicals in only a few thousand of the approximately 1,400,000 known living species. The chemicals of whole phyla of creatures have not yet received more than cursory attention. Moreover, these numbers tell only a small part of the story, because there are many more unexamined living species than they imply. We actually have only the vaguest notion regarding how many different species there really are in the world. When we say that there are so many known living species, we mean that scientists have formally described this many species and assigned Latin names to them. These enumerated species account for only a fraction of all the living species, but no one knows how large a fraction. No one seems to think there are fewer than about 4 million species, and some biologists quite seriously believe there are 30 million, 50 million, or even more! Whatever the number, there should be organisms for scientists to examine for centuries to come.

Specialists estimate that they have classified fewer than 10 percent of the total numbers of fungi, nematodes, or insects. This is intriguing, because these creatures are prolific producers of chemicals. What treasures await us in the undiscovered 90-plus percent?

Given the enormous number of undiscovered species, nature's storehouse of chemical compounds obviously remains largely unexplored. Our past ventures into this collection have brought forth such riches as lifesaving antibiotics, a beautiful fabric, and biodegradable insecticides. It is easy to predict that the search for chemicals will continue to be profitable in the years to come. In fact, the potential for profit is increasing all the time, because our options for taking advantage of natural chemicals are evolving rapidly.

Originally, when we found an interesting natural chemical, there was a single method of exploiting it: we could obtain the chemical from the organism that made it. This method gave us all our natural preparations for centuries, and it is still widely utilized today. It furnishes us with silk directly from silkworms and cocaine directly from coca leaves.

Another option for exploiting chemicals came with the advent of synthetic chemistry in the nineteenth century. After discovering a natural compound, we could synthesize it or some improved version of it in the laboratory. In the 1890s, this approach led to aspirin as an alternative to natural salicylates. At that time, salicylates were marketed to relieve pain and inflammation, and German chemists designed aspirin to compete with them. The synthetic compound was cheap and effective, and it drove the natural drugs out of the market. We now manufacture and consume thousands of tons of aspirin every year.

Until recently, these were the two principal approaches to employing natural compounds. Now, thanks to molecular biology, there is a third option, which focuses on transferring genes from one organism to another. We discovered earlier that gene-transfer techniques may eventually lead to bacteria that make spider silk. Genetic engineering has not yet given us useful silk-producing bacteria, but a related achievement in 1994 reminds us that gene-transfer techniques have already provided other products to the market.

This related achievement of gene-transfer concerns improved milk production, and it has attracted wide public attention in the United States.

Some vocal groups felt it threatened the national milk supply, while others hailed it as the first of many agricultural improvements afforded by genetic engineering. The new development grew out of the observation that dairy cows give more milk when they receive supplemental doses of their growth hormone. This hormone is a protein called bovine somatotropin (BST) that cows synthesize and use regularly in their own bodies. In February of 1994, the federal Food and Drug Administration (FDA) announced that dairy farmers would henceforth be permitted to administer supplemental BST to their cows to enhance milk production. After ten years of study, government scientists had concluded that milk from cows treated with BST was indistinguishable from other milk. (However, the BST controversy is not over. An English study appeared in October 1994 claiming that supplemental BST may harm cows.)

The commercially available BST that now boosts milk production on American dairy farms is manufactured by bacteria. The genetic instructions for synthesis of BST were isolated from bovine DNA and transferred into bacteria some years ago. Molecular biologists located the instructions in the DNA that every cow carries in its cells. They cut out the appropriate piece of bovine DNA and stitched it into the DNA of a bacterium. When each bacterial cell that carries this modified DNA divides, the BST instructions are passed on along with the rest of the cell's DNA. These bacteria make BST just as they make their own enzymes. Raising and harvesting these altered, so-called transgenic, bacteria on a large scale has made BST readily available. With bacterial BST now on the market, spider silk harvested from bacteria seems not so farfetched.

As transferring genes becomes more routine, we can expect genes themselves to become prime sources of natural chemicals. These genetic techniques permit an organism to furnish the instructions for synthesis of a chemical compound, rather than serve as the ongoing source of the compound itself. More and more chemicals will be available by this route, as biologists learn to transfer more and more extensive genetic instructions.

In the future, transferring the appropriate sequence of DNA from a rare species into bacteria will transform an unavailable chemical into a common one. Transgenic bacteria incorporating the instructions become the practical source of the chemical, and the species that furnished the gene is no longer part of the process. Once this is possible, chemicals from whatever natural source should become available without sacrificing organisms other than cheap laboratory bacteria.

Gene-transfer technology may also solve a closely related supply problem. A desirable chemical compound obtained commercially from an uncommon species may pose a dilemma at present, because great demand for the chemical threatens the continued existence of its source. This happened in the mid-nineteenth century when the demand for quinine led to the shortsighted destruction of essentially all the cinchona trees in South America. In the early 1990s, the market for the anticancer drug Taxol precipitated a similar crisis. The only known source of Taxol was the bark of the Pacific yew tree, and removing the bark to obtain the drug kills the tree. It seemed that the Pacific yew was headed for annihilation as pharmaceutical houses struggled to meet the inexhaustible demand for Taxol. If the yew disappeared, then of course there would be no more Taxol at all. The continued existence of the yew was threatened, along with the several-hundred-million-dollar annual market for Taxol.

Fortunately, innovative research provided a timely solution to this dilemma, assuring the supply of Taxol and saving the Pacific yew from extinction. Chemists developed a practical method for preparing Taxol from a close chemical relative of the drug that is found in the needles of a European yew. Needles can be harvested without killing the trees. In the future, it may be feasible to isolate the DNA that carries the genes responsible for Taxol from one yew, put these instructions into bacteria, and then raise bacteria that make Taxol.

Genetic engineering also permits biologists to transfer genes into organisms more complex than bacteria. For example, it is now frequently practical to transfer the genetic blueprint for a useful trait from one plant to another. One possibility is for agricultural scientists to take the instructions for new defensive chemicals from their plant sources and incorporate them into the DNA of crop plants, creating crops with exotic chemical defenses and improved pest resistance. In the past, moving traits from plant to plant required years of laborious breeding and hybridizing through many generations.

By creating hardier crop plants, gene-transfer technology should be a significant factor in biorational, environmentally safe pest control. What could be better than crop plants that safely generate all the pesticides they need as they grow? This is exploiting natural chemicals without ever taking them out of nature. The biotechnology for improving crop plants in this fashion is developing rapidly, and in November 1994, EPA began to gather public comments on a proposal to regulate agricultural use of these transgenic plants.

Genetic techniques will allow commercial production of natural chemicals that would otherwise remain laboratory curiosities. Both direct isolation and commercial synthesis of natural compounds will no doubt continue to be important, but for certain chemicals, such as BST and spider silk, these techniques are impractical. By providing an innovative way to manufacture chemicals, the new genetic techniques expand the pool of targets for commercial development. In view of the unexplored millions of living species, a fantastic number of compounds with beneficial properties should be within reach for sensible and safe exploitation by one means or another. It is certainly necessary to proceed responsibly and carefully, but the prospect is exhilarating. The practical results should be rewarding both socially and economically.

We could stop on this optimistic note, but before closing we should return to a topic touched upon earlier. Nature's chemicals have an opportunity to contribute to another, broader purpose in the coming years. They can help preserve the diversity of life on earth. Just as the prospect of endless benefits from natural chemicals is beginning to materialize, the natural source of these chemical riches is in decline. Only living species can supply novel natural chemicals, and the number of living species is currently decreasing at a rate unparalleled in human history. In the early twentieth century, known species became extinct at the rate of about one hundred per year, but the annual rate now is measured in the thousands. This loss in biodiversity is extremely distressing for many reasons, but even if we consider only the accompanying loss in exploitable chemical compounds, we must view it as a major disaster. Species are going extinct before being described or even discovered, and their unique chemical treasures are going with them into oblivion. Potential assets are disappearing before we learn of their existence.

The surge in species loss has several causes. It reflects expanding human populations, rising expectations, and accelerating claims on natural resources. Its single most important cause today is the active destruction of the world's tropical forests, because these forests are home to a large fraction of the undescribed living species. There are other important reservoirs of new species, but in these tropical forests, the diversity of life is truly astonishing: a single tree in a Central or South American forest may shelter more than one thousand species of beetles! Tropical forests are disappearing at a rate of 100,000 square kilometers (more than 38,000 square miles) each year. A forested area about the size of Virginia or South Korea vanishes from the tropics annually, as it is irreversibly turned into timber and cleared for development. Along with the forest, untold numbers of living species vanish as well.

We mentioned this catastrophe previously in connection with the Costa Rican National Institute of Biodiversity, and we noted the difficulty in halting the onslaught on tropical forests. Scientific arguments against deforestation stress long-term consequences, such as its disastrous effects on the climatic and biological stability of the entire earth. These long-term arguments fall on deaf ears. Moral and esthetic arguments against deforestation carry little weight with governments whose first priorities are economic development and the support of growing populations. In these circumstances, forests sometimes become nothing more than sources of quick cash and land for impoverished peasants. For its part, the logging industry generally looks to its return on investment and does whatever governments allow, or more.

In closing, it is worth mentioning again that the potential for developing chemical resources may help deter the clearing of tropical forests. In contrast to unheeded moral, esthetic, and scientific arguments, a program of deriving useful chemicals from the forests speaks to political needs and economic aspirations in developing lands. This is not the only approach to saving tropical forests, but a few valuable drugs developed from the Costa Rican experiment may work wonders. They may be more convincing to political leaders and land owners in tropical countries than any amount of debate. Key figures may then agree that harvesting drugs and other exploitable chemicals could turn their forests into renewable resources that contribute to sustainable growth.

THE CHEMICALS from nature's storehouse unite the living world in fascinating ways that we are just beginning to understand. At the same time, they offer us an unlimited source of products for human benefit. Maybe they can even help save a piece of the planet.

A NOTE TO
THE READER

*I*T MAY BE helpful to discuss briefly both the biological classification of living organisms and the scientific units of measurement employed throughout the book. Entire books devoted to each of these topics are available, and the discussion here is only a general guide.

Taxonomic Classification   Biologists specializing in an area called taxonomy, or systematics, organize living creatures into groups that have similar characteristics. Every species belongs to a genus (plural *genera*), family, order, class, phylum (plural *phyla*), and kingdom—each of which has a Latin name. Additional subdivisions, such as tribe, subclass, and superorder, are added when necessary for clarification. Occasionally, a subspecies name follows the species name. In general, several similar families belong to the same order, several similar orders belong to the same class, and so forth. Customary usage is to capitalize the name of the genus and all larger groups but not the species, and to italicize the genus and the species but not the larger groups.

For example, the house mouse (*Mus musculus*) is the species *musculus* in the genus *Mus*, which also includes about three dozen other species of mice. *Mus* and related genera belong to the family Muridae, a group of over 1,000 mice, rats, gerbils, and kindred creatures. The Muridae are one of many families that make up the order Rodentia, the rodents. (Three other families of rodents are represented by beavers, squirrels, and porcupines.) The Rodentia are one of some twenty orders of mammals, which together comprise the class Mammalia. (Other mammals include whales, bats, chimpan-

213

zees, and rabbits, each of which represents a different order.) The Mammalia, in turn, are one of several classes in the phylum Chordata, which is composed very largely of the vertebrates. (Other vertebrate classes comprise reptiles, birds, amphibians, and fishes.) Finally, the Chordata are in the kingdom Animalia, as are all other animals. (Other animal phyla include the Nematoda [round worms], Cnidaria [including corals and jellyfish], and Mollusca [mollusks].)

Until the mid-twentieth century, most biolgists were satisfied with two kingdoms: plants and animals. Modern systems of classification, however, typically divide living creatures into five kingdoms: bacteria; fungi; plants; animals; and a kingdom that includes algae, protozoa, water molds, and several other groups of creatures.

Units of Measurement   Scientific measurements are expressed using an internationally agreed-upon version of the metric system. The gram is the common unit of mass (1 gram = 0.035 ounce or 28.35 grams = 1 ounce), and the meter is the common unit of length (1 meter = 39.37 inches). The common unit of volume is the liter (1 liter = 1.057 U.S. quarts).

Prefixes are attached to indicate multiples or fractions of these units. The prefixes regularly used in this book are *centi-* (one hundredth), *milli-* (one thousandth), *micro-* (one millionth), and *nano-* (one billionth). Thus, a centimeter is one hundredth of a meter (about four tenths of an inch), and a millimeter is one thousandth of a meter (about one twenty-fifth of an inch). A microgram is one one-millionth of a gram.

When describing minute objects, we commonly refer to grains of table salt, which each measure about 200 to 500 micrometers in diameter and weigh 100 to 500 micrograms.

214

INDEX

215